C000004058

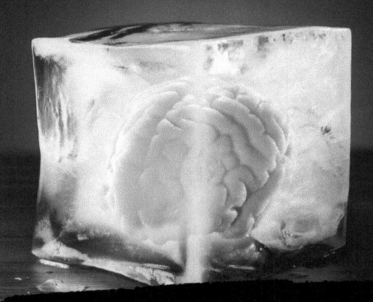

# CRYONICS

The Science of Life Extension

## BY ETHAN D. ANDERSON

WWW.BORNINCREDIBLE.COM

# Table of Contents

**Get A Free Book At: BornIncredible.com/free-book-offer/[4]**

BORNINCREDIBLE

5

---

1. https://BornIncredible.com

2. https://BornIncredible.com

3. https://BornIncredible.com

**4. https://bornincredible.com/free-book-offer/**

5. https://bornincredible.com/

# The History of Cryonics

The concept of cryonics, the preservation of the human body or brain at extremely low temperatures with the hope of future revival and medical intervention, has captured the imagination of many. It represents a fascinating intersection between science, technology, and the quest for immortality. The history of cryonics is a testament to the human desire to conquer the limitations of mortality and explore the possibilities of life extension.

The roots of cryonics can be traced back to the mid-20th century, with the pioneering work of Robert Ettinger, an American physics teacher. In 1962, Ettinger published "The Prospect of Immortality," a book that outlined the idea of cryonic suspension as a means to preserve human bodies after death. He proposed that by subjecting the body to extremely low temperatures, cellular deterioration and decay could be halted, providing an opportunity for future medical advancements to revive and restore life.

Ettinger's book sparked widespread interest and ignited discussions among scientists, philosophers, and the general public. It laid the foundation for the establishment of the first cryonics organization, the Cryonics Society of Michigan, in 1965. This organization aimed to offer cryonic preservation services to individuals who desired to have their bodies or brains stored in a suspended state after death.

In the following years, cryonics gained both supporters and skeptics. Cryonicists believed that future advancements in medical science and technology, such as nanotechnology and tissue regeneration, could potentially reverse the aging process and restore life to those in cryonic suspension. They saw cryonics as a form of "life insurance" that offered a chance for future revival and the opportunity to continue living.

However, cryonics also faced significant challenges and criticism. Skeptics questioned the feasibility and ethical implications of cryonic preservation. Many argued that the process of freezing and thawing could cause irreparable damage to cells and tissues, making revival and restoration highly unlikely. Others raised ethical concerns, questioning the moral implications of extending life indefinitely and the potential inequality that cryonics could create if only a select few could afford the procedure.

Despite the skepticism, cryonics continued to evolve and develop as a field. Cryonics organizations emerged in various parts of the world, offering preservation services and refining the techniques and protocols for cryonic suspension. Advances in cryopreservation techniques, such as the introduction of vitrification, which reduces the formation of ice crystals during the freezing process, have shown promising results in minimizing cell damage.

The first human to be cryonically preserved was Dr. James Bedford, who passed away in 1967 and was placed in cryonic suspension. Since then, hundreds of individuals have chosen cryonics as an option for post-mortem preservation. Cryonics organizations provide storage facilities where bodies or brains are maintained at temperatures below -196 degrees Celsius, typically in liquid nitrogen.

In recent years, cryonics has gained attention from the scientific community, with researchers exploring the principles of cryobiology and the potential applications of cryopreservation techniques. While the focus has primarily been on the preservation of organs and tissues for transplantation, the knowledge gained from these studies can contribute to the understanding and improvement of cryonic suspension techniques.

It is important to note that cryonics remains an experimental and controversial field. The technology and scientific understanding necessary to revive a cryonically preserved individual are still beyond our current capabilities. However, proponents argue that cryonics is a form of "last hope" for individuals who have no other options and believe in the potential of future advancements in science and medicine.

The history of cryonics is a testament to the human fascination with immortality and the desire to explore the boundaries of life and death. It has sparked debates about the nature of consciousness and the ethical implications of extending life indefinitely.

# Early Experiments and Pioneers

The field of cryonics, which involves the preservation of the human body or brain at extremely low temperatures, has its roots in a series of early experiments and the pioneering work of several individuals. These early pioneers laid the foundation for the development of cryonics as we know it today.

One of the earliest experiments related to cryonics can be traced back to the late 18th century when scientists began exploring the effects of low temperatures on living organisms. The Italian scientist Giovanni Aldini conducted experiments on animal tissues and organs, demonstrating the potential of cold temperatures to slow down biological processes. Aldini's work set the stage for future investigations into the field of cryonics.

1. In the early 20th century, a prominent figure in the field of cryonics emerged: Dr. James Lovelock. Lovelock, an English scientist, conducted experiments in the 1930s that involved freezing and reviving simple organisms. His research laid the groundwork for the idea that low temperatures could preserve life and potentially enable the revival of organisms.

Another significant figure in the early days of cryonics was Dr. Robert C. W. Ettinger, an American physics teacher. Ettinger published the influential book "The Prospect of Immortality" in 1962, which discussed the concept of cryonic suspension as a means to preserve human bodies after death. He proposed that by subjecting the body to extremely low temperatures, cellular deterioration and decay could be halted, providing an opportunity for future medical advancements to revive and restore life.

Ettinger's book garnered attention and inspired the formation of the first cryonics organization, the Cryonics Society of Michigan, in 1965. The society aimed to offer cryonic preservation services to individuals who wished to be preserved in a suspended state after death. This marked the beginning of organized cryonics efforts and the establishment of a community of individuals dedicated to the exploration of life extension through cryonics.

One of the early pioneers who contributed significantly to the development of cryonics techniques was Curtis Henderson. Henderson, a biophysicist, developed the concept of cryonic suspension using glycerol as a cryoprotective agent. His experiments showed that glycerol could protect biological tissues from damage caused by freezing and thawing. Henderson's work paved the way for improved cryopreservation techniques and laid the foundation for the use of cryoprotective agents in preserving organs and tissues.

Over time, cryonics organizations such as the Cryonics Institute and Alcor Life Extension Foundation emerged, refining the techniques and protocols for cryonic suspension. These organizations conducted research, developed storage facilities, and provided cryonic preservation services to individuals who opted for this method of post-mortem preservation.

The pioneers of cryonics faced significant challenges and skepticism from the scientific and medical communities. Many scientists questioned the feasibility and scientific basis of cryonics, highlighting the potential damage that could occur during the freezing and thawing process. Ethical concerns were also raised, questioning the implications of extending life indefinitely and the potential inequality that cryonics could create if only a select few could afford the procedure.

Despite the challenges, early pioneers of cryonics remained dedicated to their vision of preserving life and exploring the possibilities of future revival. They continued to refine cryopreservation techniques, explore new cryoprotective agents, and advocate for the scientific study of cryonics.

Today, cryonics remains a controversial field, with proponents arguing for its potential as a last resort for individuals who have no other options and believe in the future advancements of science and medicine. Research and technological advancements continue to be made in the field, with scientists exploring the principles of cryobiology and the potential applications of cryopreservation techniques.

# The Birth of Modern Cryonics

The birth of modern cryonics can be traced back to the mid-20th century when several key developments and advancements took place, paving the way for the establishment of organized cryonics efforts. This period marked a turning point in the history of cryonics, as it transitioned from theoretical concepts to practical application.

One of the pivotal moments in the birth of modern cryonics was the publication of the book "The Prospect of Immortality" by Dr. Robert C. W. Ettinger in 1962. Ettinger's book introduced the idea of cryonic suspension as a means to preserve human bodies after death. He proposed that by subjecting the body to extremely low temperatures, cellular deterioration and decay could be halted, allowing for the possibility of future revival and restoration of life.

"The Prospect of Immortality" sparked widespread interest and ignited discussions among scientists, philosophers, and the general public. It laid the foundation for the establishment of the first cryonics organization, the Cryonics Society of Michigan, in 1965. This organization aimed to provide cryonic preservation services to individuals who desired to have their bodies or brains stored in a suspended state after death.

The early years of organized cryonics efforts were marked by experimentation and the development of preservation techniques. One of the key figures during this time was Dr. James H. Bedford, who became the first person to be cryonically preserved in 1967. Bedford's body was placed in cryonic suspension with the hope that future advancements in science and technology would allow for his revival.

The establishment of cryonics organizations, such as the Cryonics Institute and Alcor Life Extension Foundation, in the 1970s further solidified the infrastructure for cryonic preservation. These organizations offered storage facilities where bodies or brains could be maintained at temperatures below -196 degrees Celsius, typically in liquid nitrogen. They developed protocols and procedures for cryonic suspension, including the use of cryoprotective agents to minimize cellular damage during the freezing process.

During this period, cryonics also faced significant challenges and skepticism from the scientific and medical communities. Many scientists questioned the feasibility and scientific basis of cryonics, raising concerns about the potential damage that could occur during the freezing and thawing process. Ethical considerations were also debated, with discussions revolving around the nature of consciousness, the definition of death, and the potential inequalities that cryonics could create.

Despite the challenges, the birth of modern cryonics brought together a community of individuals dedicated to the exploration of life extension through cryonics. Cryonicists believed in the potential of future advancements in science and technology to reverse the aging process and restore life to those in cryonic suspension. They saw cryonics as a form of "life insurance," offering a chance for future revival and the opportunity to continue living.

Over the years, cryonics has continued to evolve and develop as a field. Advances in cryopreservation techniques, such as the introduction of vitrification, have shown promise in minimizing cell damage during the freezing and thawing process. Vitrification involves the use of cryoprotective solutions to transform tissues into a glass-like state, reducing the formation of ice crystals that can cause cellular damage.

The birth of modern cryonics also sparked scientific research and technological advancements related to cryobiology and cryopreservation. Researchers have explored the principles of cryopreservation and the effects of low temperatures on biological tissues, laying the groundwork for future improvements in preservation techniques.

Today, cryonics remains a controversial and speculative field. It continues to inspire debate and discussion among scientists, philosophers, and the general public. While cryonic suspension techniques have improved, the technology and scientific understanding necessary to revive a cryonically preserved individual are still beyond our current capabilities.

# Key Developments and Milestones

Key developments and milestones have played a crucial role in shaping the field of cryonics, which focuses on the preservation of the human body or brain at extremely low temperatures with the hope of future revival. These milestones reflect the progress made in both the scientific and practical aspects of cryonics, advancing our understanding and capabilities in this fascinating area of research.

One significant milestone in the history of cryonics was the establishment of the Cryonics Society of California in 1969, which later became the Alcor Life Extension Foundation. Alcor played a pivotal role in developing the infrastructure and protocols for cryonics, including the establishment of cryopreservation facilities and the refinement of preservation techniques. Alcor's contributions have significantly influenced the field and facilitated the growth of cryonics as a scientific and practical endeavor.

Another notable development in cryonics was the introduction of vitrification, a preservation method that minimizes ice crystal formation during the freezing process. In the late 1980s, cryobiologist Greg Fahy developed a technique called cryoprotectant vitrification, which involves replacing water inside cells with a cryoprotective agent that solidifies into a glass-like state upon cooling. This method has shown promise in reducing cellular damage associated with traditional freezing methods, potentially improving the chances of successful revival in the future.

The year 1991 marked a significant milestone in the field of cryonics with the cryopreservation of the first human whole body by Alcor. This landmark event demonstrated the practical application of cryonics on a full-scale basis, reinforcing the notion that cryonic preservation was not limited to experimental animals or isolated tissue samples but extended to human individuals. It provided a foundation for further advancements and inspired continued research and development in the field.

Advances in cryopreservation techniques and technologies have continued to drive progress in cryonics. In 2005, a breakthrough was achieved with the successful vitrification and cryopreservation of a rabbit kidney by a team led by Fahy. This achievement demonstrated that the vitrification technique could be applied to complex organs and tissues, bringing us closer to the possibility of preserving more complex biological systems.

Another significant development in recent years has been the emergence of neurocryopreservation as an alternative option in cryonics. Neurocryopreservation involves preserving only the brain rather than the whole body, with the belief that the brain contains the essential information for future revival. This approach is seen as a potentially more efficient and cost-effective method of cryonic preservation.

In addition to advancements in preservation techniques, there have been notable strides in the scientific understanding of cryobiology and the effects of low temperatures on biological systems. Cryobiologists have conducted research to investigate the mechanisms of cryopreservation and to identify strategies to minimize damage to cells and tissues during the freezing and thawing processes. This knowledge is crucial for improving cryonic techniques and increasing the chances of successful revival in the future.

The milestones achieved in cryonics have not been without challenges and skepticism. Cryonics has faced criticism from some sectors of the scientific community, who question the feasibility and ethical implications of preserving human bodies or brains in a state of suspended animation. The issues of reversibility, cellular damage, and the complex nature of revival remain significant hurdles to overcome.

Despite the challenges, cryonics continues to evolve and generate interest from scientists, futurists, and those seeking alternative approaches to life extension. The field has inspired discussions on the nature of consciousness, the possibilities of future technologies, and the boundaries of human existence. While the realization of successful revival from cryonic preservation remains speculative, the field continues to attract individuals who see it as a potential pathway to future advancements in science and medicine.

# Cryonics: The Science Behind It

Cryonics, the practice of preserving the human body or brain at extremely low temperatures, has captivated the imaginations of many. It represents a unique intersection of science, technology, and the pursuit of life extension. But what is the science behind cryonics? How does it work? Below we will explore the scientific principles and techniques that underpin the field of cryonics.

At its core, cryonics is based on the principle that extremely low temperatures can halt biological processes and preserve the structure of cells and tissues. The goal is to suspend the body or brain in a state of cryonic preservation, with the hope that future advancements in science and technology will allow for revival and restoration of life.

One of the fundamental challenges in cryonics is the preservation of cellular integrity during the freezing and thawing processes. The formation of ice crystals can cause significant damage to cells, disrupting their structure and function. To address this, cryonicists have developed techniques such as vitrification, which aims to minimize ice crystal formation and preserve cellular structures in a glass-like state.

Vitrification involves the use of cryoprotective agents that penetrate the cells and replace the water molecules. These agents prevent ice crystal formation by solidifying into a glass-like state at low temperatures. By vitrifying tissues, cryonicists aim to maintain the structural integrity of cells and minimize the damage caused by ice crystals.

The process of vitrification begins with a series of steps to prepare the body or brain for cryonic preservation. The individual is cooled down gradually to a temperature around -130 degrees Celsius using a technique called perfusion. During perfusion, cryoprotective agents are infused into the circulatory system to replace the blood and other fluids in the body. This step helps to prevent ice crystal formation and allows for better preservation of the tissues.

Once the cryoprotective agents have been circulated throughout the body, the individual is further cooled down to a temperature below -196 degrees Celsius, at which point they are transferred to long-term storage in liquid nitrogen. Liquid nitrogen is an ideal medium for long-term preservation, as it maintains a stable temperature and slows down the metabolic processes that could lead to degradation.

While the science behind vitrification and cryopreservation techniques has made significant progress, there are still challenges to overcome. Reversing the process and restoring life to a cryonically preserved individual remains a complex and speculative endeavor. The revival process would require advances in medical and technological fields that are yet to be realized.

One of the key challenges in cryonics is the issue of tissue damage caused by the cryoprotective agents themselves. Although cryoprotective agents are designed to protect cells, their presence can still have an impact on the tissues. Cryonics organizations continue to research and develop improved cryoprotective agents that minimize potential damage and maximize preservation quality.

Another challenge is the need for future technological advancements in revival techniques. Cryonics is closely tied to other scientific and medical disciplines, such as nanotechnology and regenerative medicine. The hope is that these fields will progress to a point where they can repair and restore damaged tissues, reverse the aging process, and eventually revive cryonically preserved individuals.

While the scientific basis for cryonics is founded on principles of cellular preservation and the prevention of ice crystal formation, it is important to note that cryonics remains a speculative field. The scientific community at large remains cautious and skeptical about the feasibility of revival and the potential for restoring function to cryonically preserved individuals.

Nevertheless, cryonics continues to inspire scientific exploration and research. It raises questions about the nature of consciousness, the boundaries of life and death, and the potential of future technologies.

# An Overview of the Cryopreservation Process

Cryopreservation, the process of preserving biological materials at very low temperatures, is a critical aspect of cryonics, the practice of preserving the human body or brain in the hope of future revival. Below we will provide an overview of the cryopreservation process, highlighting the key steps and techniques involved.

The cryopreservation process begins immediately after legal death, with the goal of minimizing cellular damage and preserving the structure and integrity of tissues for long-term storage. The process involves several important stages, including stabilization, perfusion, cooling, and storage.

The first step in cryopreservation is stabilization. To prevent immediate deterioration after legal death, the body or brain is rapidly cooled to a temperature just above freezing using cooling blankets or ice packs. This step helps slow down metabolic processes and delays cellular degradation.

Following stabilization, the cryopreservation process moves into the perfusion phase. Perfusion involves the circulation of cryoprotective agents through the vascular system to protect cells from the damaging effects of freezing. Cryoprotective agents are chemical solutions that penetrate cells, replacing water and reducing ice formation during the freezing process.

Perfusion is typically performed by accessing major blood vessels, such as the carotid artery and jugular vein, in the case of whole-body cryopreservation, or the femoral artery and vein, for brain-only preservation. Cryoprotective agents are then slowly introduced into the circulatory system while simultaneously removing blood and other fluids from the body or brain.

The choice of cryoprotective agents is crucial to the success of cryopreservation. Commonly used cryoprotectants include glycerol, dimethyl sulfoxide (DMSO), and ethylene glycol. These agents have the ability to vitrify or transform into a glass-like state at low temperatures, reducing the formation of ice crystals and minimizing cell damage.

Once perfusion is complete and cryoprotective agents have circulated throughout the tissues, the cooling process begins. This phase involves gradually reducing the temperature of the body or brain to the storage temperature of liquid nitrogen (-196 degrees Celsius or -320.8 degrees Fahrenheit).

The rate of cooling is critical to prevent the formation of ice crystals and minimize cellular damage. Slow cooling techniques, such as computer-controlled cooling or cooling chambers, are used to maintain a controlled and uniform rate of cooling. This allows for optimal preservation of the tissues and minimizes the potential for structural damage.

Once the desired temperature is reached, the body or brain is placed in a storage container filled with liquid nitrogen. Liquid nitrogen provides a stable and extremely low-temperature environment for long-term preservation. It prevents further degradation and slows down metabolic processes, effectively putting the biological material in a state of suspended animation.

Storage containers used in cryonics are designed to maintain a consistent temperature and ensure the integrity of the cryopreserved tissues. They are often made of materials with low thermal conductivity, such as stainless steel or high-quality plastics, to minimize heat transfer and maintain the desired temperature.

Regular maintenance and monitoring of the storage facilities are essential to ensure the integrity of the cryopreserved samples. Cryonics organizations have stringent protocols in place to maintain the stability and security of the storage containers, including periodic checks of liquid nitrogen levels, temperature monitoring, and security measures to prevent unauthorized access.

It is important to note that cryopreservation is a relatively new and evolving field, and there are ongoing research efforts to improve techniques and develop more efficient cryoprotective agents. The ultimate goal is to achieve reversible cryopreservation, where cells and tissues can be successfully thawed and restored to their original state without significant damage.

# Vitrification: A Breakthrough in Cryopreservation

Vitrification, a breakthrough in cryopreservation, has revolutionized the field of cryonics by addressing one of its fundamental challenges: the formation of ice crystals during freezing. This innovative technique has shown promise in preserving biological materials, such as cells and tissues, with minimal damage, opening up new possibilities for the preservation and potential revival of living organisms.

Traditional cryopreservation methods involve the use of slow cooling rates to minimize ice crystal formation. However, even with controlled cooling, ice crystals can still form and cause significant damage to cellular structures. Vitrification, on the other hand, aims to eliminate ice crystal formation altogether by solidifying the biological material into a glass-like state.

The vitrification process involves the use of high concentrations of cryoprotective agents, such as ethylene glycol or dimethyl sulfoxide (DMSO), which have the ability to penetrate cells and prevent ice crystal formation. These cryoprotectants act as antifreeze agents, reducing the freezing point of water and altering its properties.

To achieve vitrification, the biological material is exposed to a cryoprotective solution that contains a high concentration of cryoprotectants. The concentration of cryoprotectants is critical to the success of vitrification, as it determines the ability of the solution to vitrify and solidify into a glass-like state at low temperatures.

Once the biological material has been immersed in the cryoprotective solution, it is rapidly cooled to ultra-low temperatures using techniques such as direct plunging into liquid nitrogen or ultra-fast cooling methods. The rapid cooling rate prevents the formation of ice crystals and allows the cryoprotective solution to solidify, transforming the biological material into a glass-like state.

In this vitrified state, the cellular structures and components are immobilized, maintaining their original positions and preserving their integrity. Unlike in traditional cryopreservation methods, where ice crystals can cause physical damage to cells, vitrification ensures that the biological material remains structurally intact, minimizing the potential for damage during the freezing and thawing processes.

One of the significant advantages of vitrification is its potential for ultra-rapid cooling and rewarming. With traditional slow-cooling methods, the process of cooling and rewarming can be time-consuming, increasing the risk of cellular damage. Vitrification, on the other hand, allows for rapid cooling and rewarming, which reduces the exposure time to potentially damaging conditions.

However, vitrification also presents challenges. The high concentrations of cryoprotectants required for vitrification can be toxic to cells and tissues. To mitigate this, careful optimization of the vitrification solutions and protocols is necessary to minimize toxicity while maintaining vitrification efficiency.

Another challenge is the prevention of crystallization during rewarming. The transformation of the vitrified material back into a liquid state without the formation of ice crystals is crucial to preserve the structural integrity of the biological material. Innovative techniques, such as nanotechnology and advanced warming methods, are being explored to overcome this challenge and improve the rewarming process.

Vitrification has found applications not only in cryonics but also in other fields, such as regenerative medicine and assisted reproductive technologies. In these fields, vitrification has been used to preserve embryos, oocytes, and tissues, enabling the storage and future use of these biological materials.

The successful application of vitrification in cryonics has paved the way for advancements in the preservation of whole organs and complex tissues. Research efforts are underway to refine vitrification protocols and explore its potential for preserving more complex biological systems.

While vitrification represents a significant breakthrough in cryopreservation, it is important to note that the process of revival and restoration of vitrified biological materials is still a challenging and speculative endeavor.

# The Role of Cryoprotectants

Cryoprotectants play a crucial role in the field of cryonics, where the preservation of biological materials at extremely low temperatures is key. These specialized compounds have unique properties that help protect cells and tissues from damage during the freezing and thawing processes. Below we will explore the role of cryoprotectants in cryonics and their significance in preserving biological materials.

The primary function of cryoprotectants is to prevent ice crystal formation during freezing. When a biological material is cooled, the water inside the cells can form ice crystals, which can cause physical damage to the cellular structures. Cryoprotectants work by reducing the freezing point of water and altering its properties, thus minimizing ice crystal formation.

There are various types of cryoprotectants used in cryonics, including dimethyl sulfoxide (DMSO), ethylene glycol, propylene glycol, and glycerol. These compounds have the ability to penetrate the cell membranes and interact with the water molecules inside the cells, preventing the formation of ice crystals.

One commonly used cryoprotectant is dimethyl sulfoxide (DMSO), a versatile compound with excellent membrane penetration properties. DMSO is known for its ability to reduce the freezing point of water, making it an effective cryoprotectant. It has been extensively used in cryonics to protect cells and tissues during the freezing process.

Ethylene glycol and propylene glycol are other commonly used cryoprotectants. They are similar in structure to glycerol and can also reduce the freezing point of water. These cryoprotectants are often used in combination with other agents to optimize the cryopreservation process.

Glycerol, a natural compound, has been used as a cryoprotectant for many years. It is known for its ability to penetrate cell membranes and act as a protective agent during freezing. Glycerol is particularly effective in protecting red blood cells and sperm cells during cryopreservation.

The choice of cryoprotectant depends on several factors, including the type of biological material being preserved and the specific goals of the cryopreservation process. Different cryoprotectants have different properties and may be more suitable for certain applications.

In cryonics, cryoprotectants are often used in high concentrations to achieve vitrification, a state where the biological material solidifies into a glass-like structure instead of forming ice crystals. Vitrification minimizes cellular damage during freezing and allows for better preservation of the structure and integrity of the biological material.

The concentration of cryoprotectants used in cryonics is carefully determined to balance the need for effective protection with the potential for toxicity. Cryoprotectants, especially at high concentrations, can be toxic to cells and tissues. Finding the optimal concentration is crucial to ensure maximum protection while minimizing adverse effects.

To further enhance the efficiency of cryoprotectants, multi-step perfusion techniques are often employed. Perfusing the biological material with progressively increasing concentrations of cryoprotectants allows for better penetration and distribution within the tissues, ensuring uniform protection during the freezing process.

It is important to note that cryoprotectants are not a universal solution, and the choice of cryoprotectant and its concentration may vary depending on the specific biological material being preserved. Research and experimentation continue to improve cryoprotectant formulations and protocols to enhance cryopreservation outcomes.

While cryoprotectants have significantly improved the field of cryonics, challenges still exist. The potential toxicity of cryoprotectants and their long-term effects on cellular structures require careful consideration.

# The Ethics and Controversies of Cryonics

Cryonics, the practice of preserving the human body or brain at extremely low temperatures with the hope of future revival, raises a multitude of ethical and controversial questions. The pursuit of life extension and the preservation of consciousness beyond death prompt debates surrounding the scientific feasibility, moral implications, and potential inequalities associated with cryonics. Below we will explore the ethical and controversial aspects surrounding cryonics.

One of the primary ethical considerations in cryonics revolves around the concept of informed consent. Cryonic preservation requires individuals to make decisions about their own bodies and the fate of their remains after death. Critics argue that cryonics may exploit vulnerable individuals who might be influenced by false promises of future revival. They raise concerns about the level of understanding and consent individuals have regarding the long-term consequences and scientific uncertainties associated with cryonics.

The allocation of resources is another ethical dilemma posed by cryonics. Cryonics is an expensive endeavor, with preservation and storage costs ranging from tens of thousands to hundreds of thousands of dollars. Critics argue that the significant financial investment required for cryonics perpetuates inequalities, as access to this potentially life-extending technology is limited to those with significant financial means. This raises questions about fairness and the potential for cryonics to exacerbate social disparities.

The definition of death and the timing of cryonic preservation are additional ethical challenges. Cryonics organizations require prompt preservation after legal death to maximize the chances of successful preservation. However, the determination of legal death does not necessarily align with the cessation of all brain activity. Some argue that cryonics blurs the line between life and death, as individuals may be preserved in a state where their consciousness and potential for revival are uncertain. Ethical questions arise as to whether cryonic preservation should be performed before biological death or how long after legal death it should be allowed.

The potential for revival and restoration of consciousness is a central ethical concern in cryonics. Critics question the feasibility of future technologies that could revive cryonically preserved individuals. They argue that the scientific understanding and technological capabilities necessary for successful revival may never be achieved. The ethical implications of potentially reviving individuals in a future society with different values, social structures, and relationships also raise complex questions.

Cryonics also presents challenges in terms of intergenerational justice. Cryonically preserved individuals are essentially relying on future generations to develop the technologies and medical advancements necessary for revival. This raises questions about the moral responsibility of current generations in preserving resources and knowledge for future generations. Critics argue that cryonics places an undue burden on future societies, diverting resources and attention from current pressing issues such as healthcare, poverty, and environmental sustainability.

Religious and cultural perspectives add further layers of controversy to cryonics. Different belief systems have varying views on death, the afterlife, and the preservation of the body after death. Some religious traditions view cryonics as interfering with the natural order of life and death or as a form of hubris. The clash between religious and cultural beliefs and the aspirations of cryonics creates ethical dilemmas and potential conflicts.

The scientific community itself remains divided on the topic of cryonics. Skeptics argue that the scientific basis and feasibility of cryonics are unproven. They question the potential for successful revival and argue that the preservation techniques may cause irreparable damage to cellular structures. Critics also contend that cryonics diverts resources and attention from other areas of scientific research that could have a more immediate impact on improving human health and well-being.

Despite the ethical concerns and controversies surrounding cryonics, proponents argue that cryonics offers a last hope for individuals who see it as a chance for future advancements in science and medicine. They believe in the potential of future technologies to reverse the aging process, cure diseases, and restore cells.

# Philosophical Debates: Life, Death, and Identity

The field of cryonics, with its promise of preserving life beyond death, raises profound philosophical questions about the nature of life, death, and personal identity. These debates delve into the fundamental aspects of human existence, challenging our understanding of consciousness, personal continuity, and the boundaries of identity. Below we will explore the philosophical debates surrounding cryonics and its implications for our concepts of life, death, and identity.

One of the central philosophical questions raised by cryonics is the nature of life itself. Cryonics challenges traditional notions of death as the irreversible cessation of all biological processes. Advocates argue that cryonically preserved individuals are not truly dead but in a state of suspended animation, with the potential for future revival. This challenges our understanding of life as a continuous process and raises questions about the definition and boundaries of life itself.

The concept of personal identity is also at the core of philosophical debates surrounding cryonics. Cryonics raises questions about what makes us who we are and whether our personal identity can persist through the process of cryonic preservation. Critics argue that cryonics disrupts personal continuity, as the revival of a cryonically preserved individual may result in a different consciousness or a break in memory continuity. This raises questions about whether the revived individual would be the same person as the one who was originally cryopreserved.

The mind-body problem, a longstanding debate in philosophy, intersects with cryonics in profound ways. Cryonics poses the question of the relationship between the mind and the body and whether consciousness can be preserved or restored. Some argue that consciousness is an emergent property of the physical brain, and preserving the brain through cryonics offers the possibility of retaining or reviving consciousness. Others contend that consciousness is not solely dependent on physical substrates and may have non-physical aspects that transcend the body.

Cryonics also touches on the concept of personal autonomy and the right to determine the fate of one's own body. Proponents argue that cryonics provides individuals with the opportunity to exercise autonomy over their own lives and bodies, even beyond death. They contend that individuals should have the freedom to choose cryonic preservation as a means of extending their lives or exploring the possibility of future revival. However, critics raise concerns about the potential exploitation of vulnerable individuals and the validity of consent in making decisions about cryonic preservation.

The philosophical debates surrounding cryonics also encompass questions about the nature of personal experience and subjective reality. Critics argue that cryonic preservation may result in an interruption of subjective experience, as individuals would be deprived of sensory input and interaction with the external world. They question whether a preserved individual, even if revived in the future, would have a meaningful and coherent experience of life.

Cryonics also raises ethical questions about the allocation of resources and the distribution of opportunities for life extension. Critics argue that cryonics diverts resources and attention from pressing issues, such as improving healthcare, alleviating poverty, and addressing social inequalities. They raise concerns about the potential for cryonics to exacerbate existing disparities and create a scenario where only the wealthy few have access to life-extending technologies.

The philosophical debates surrounding cryonics extend to considerations of societal values and cultural beliefs. Cryonics challenges cultural and religious perspectives on death and the afterlife, raising questions about the compatibility of cryonics with different belief systems. The clash between scientific aspirations and cultural or religious traditions highlights the need for respectful dialogue and understanding in navigating the ethical and philosophical dimensions of cryonics.

# Legal and Regulatory Challenges

The practice of cryonics, which involves the preservation of the human body or brain at extremely low temperatures with the hope of future revival, presents a unique set of legal and regulatory challenges. The novel nature of cryonics raises questions regarding the legal status of cryopreserved individuals, the responsibilities of cryonics organizations, and the potential impact on existing legal frameworks. Below we will explore the legal and regulatory challenges associated with cryonics.

One of the primary legal challenges in cryonics pertains to the status of cryopreserved individuals. Legal systems typically define death as the irreversible cessation of all brain activity or as the irreversible cessation of circulation and respiratory functions. Cryonics blurs the line between life and death, as cryopreserved individuals are in a state of suspended animation rather than experiencing the traditional cessation of bodily functions. This raises questions about the legal rights and obligations of cryopreserved individuals and their potential status in the eyes of the law.

Cryonics organizations face regulatory challenges related to licensing, oversight, and accountability. The operation of cryonics facilities involves complex medical and scientific procedures, necessitating adherence to appropriate standards and protocols. However, the regulatory frameworks for cryonics may be inadequate or nonexistent in many jurisdictions, as they were not designed with this unique practice in mind. Establishing appropriate regulatory mechanisms and oversight for cryonics organizations is a significant challenge that requires careful consideration of the ethical, legal, and scientific dimensions involved.

The preservation and storage of cryonically preserved bodies or brains also raise legal questions. Cryonics organizations must navigate legal frameworks related to property rights, including the ownership and disposition of cryopreserved remains. This is particularly relevant in cases where individuals have made arrangements for cryonic preservation in their wills or estate plans. The legal status of cryopreserved remains and the recognition of individuals' wishes in their posthumous arrangements can vary across jurisdictions and may require specific legal frameworks to address these unique circumstances.

Cryonics also raises issues regarding the liability and responsibility of cryonics organizations. The preservation and storage of human remains involve potential risks and obligations. Cryonics organizations must take appropriate measures to ensure the safety and security of the cryopreserved individuals and their associated personal effects. This includes addressing concerns related to potential equipment failures, breaches of security, or mishandling of cryopreserved remains. Developing legal frameworks that address the responsibilities and liabilities of cryonics organizations is crucial to protect the interests of the individuals involved and to ensure the ethical and responsible practice of cryonics.

The international nature of cryonics further complicates the legal landscape. Cryonics organizations may operate in one jurisdiction while storing cryopreserved remains in another. This raises questions about the recognition of legal arrangements and the transfer of human remains across borders. Harmonizing legal frameworks and establishing mechanisms for international cooperation are essential for facilitating the practice of cryonics and ensuring consistent legal treatment across jurisdictions.

Another legal challenge is the recognition of cryonics in relation to existing legal frameworks, such as those pertaining to inheritance, organ transplantation, and end-of-life decision-making. Cryonics blurs the lines between life and death, challenging the traditional legal definitions and frameworks in these areas. For example, the legal recognition of cryonic preservation in inheritance laws may require specific provisions to address the unique circumstances of cryopreserved individuals. Similarly, legal frameworks surrounding organ transplantation may need to consider the potential availability of cryonically preserved organs for future transplantation.

Ethical considerations also play a crucial role in the legal and regulatory challenges of cryonics. Balancing individual autonomy and the protection of vulnerable individuals is a significant ethical concern. Legal frameworks must provide safeguards to ensure that individuals are making informed decisions and that their rights and interests are protected.

## Addressing Ethical Concerns

Cryonics, the practice of preserving the human body or brain at extremely low temperatures with the hope of future revival, raises significant ethical concerns. These concerns revolve around informed consent, resource allocation, intergenerational justice, societal values, and the preservation of personal identity. Below we will explore how these ethical concerns can be addressed within the context of cryonics.

One of the primary ethical concerns in cryonics is informed consent. It is essential to ensure that individuals who choose cryonic preservation have a clear understanding of the risks, uncertainties, and potential outcomes associated with the practice. Cryonics organizations should provide comprehensive information, education, and counseling to individuals considering cryonic preservation, allowing them to make informed decisions about their own bodies and the fate of their remains after death. This includes discussing the scientific basis of cryonics, the limitations and uncertainties surrounding revival, and the potential impact on personal identity and relationships.

Addressing resource allocation concerns is another ethical consideration in cryonics. Cryonic preservation is an expensive endeavor, and critics argue that it may exacerbate social inequalities, as access to this potentially life-extending technology is limited to those with significant financial means. To address this concern, cryonics organizations can explore mechanisms for financial assistance or alternative funding models to ensure greater accessibility. Collaboration with the scientific and medical communities, as well as efforts to promote public awareness and support, can also contribute to the development of more equitable approaches to cryonic preservation.

Intergenerational justice is an ethical concern related to cryonics, as cryonically preserved individuals rely on future generations to develop the technologies and medical advancements necessary for revival. To address this concern, cryonics organizations can promote responsible resource management and knowledge preservation. By actively contributing to scientific research and education, cryonics organizations can ensure that they are not placing an undue burden on future societies but rather actively contributing to the progress and well-being of future generations.

The clash between societal values and cryonics presents another ethical challenge. Cryonics challenges cultural and religious perspectives on death and the afterlife, potentially creating conflicts and misunderstandings. Dialogue and mutual understanding between cryonics organizations and religious or cultural institutions can help address these concerns. Open discussions about the philosophical, ethical, and scientific aspects of cryonics can foster greater acceptance and respect for diverse perspectives while facilitating an informed and nuanced approach to the practice.

Preserving personal identity is a significant ethical concern in cryonics. Critics argue that cryonics disrupts personal continuity, as revival may result in a different consciousness or a break in memory continuity. Cryonics organizations can address this concern by engaging in ongoing research and dialogue on the nature of personal identity, consciousness, and memory. This includes exploring advancements in fields such as neuroscience and psychology to better understand the potential impacts of cryonic preservation on personal identity. Additionally, cryonics organizations can develop protocols and technologies that aim to minimize disruption to personal continuity during the revival process.

Transparency, accountability, and adherence to ethical guidelines are essential for addressing the ethical concerns surrounding cryonics. Cryonics organizations can establish professional standards, codes of conduct, and ethical review boards to ensure responsible practices. Collaborating with the scientific and medical communities, as well as engaging in open dialogue with regulatory bodies, can help establish effective oversight and governance mechanisms.

Education and public engagement play a crucial role in addressing ethical concerns surrounding cryonics. Raising awareness about the scientific basis, potential benefits, and ethical considerations of cryonics can foster informed discussions and informed decision-making. Public engagement efforts, including public forums, educational initiatives, and partnerships with academic institutions, can promote a better understanding of cryonics and its ethical implications.

# Cryonics Procedures: From Life to Suspended Animation

Cryonics, the practice of preserving the human body or brain at ultra-low temperatures, involves a series of intricate procedures that transition a person from the state of life to suspended animation. These procedures are designed to minimize cellular damage and maintain the structural integrity of tissues during cryopreservation. Below we will explore the step-by-step process involved in cryonics, from the initial stages of pronouncement to the final stage of long-term storage.

The cryonics process begins with the pronouncement of legal death. Cryonics organizations must be notified promptly after legal death to initiate the cryopreservation procedures. Time is of the essence, as rapid intervention allows for a higher likelihood of successful preservation.

Upon notification, cryonics teams mobilize to stabilize the patient's body or brain. This involves cooling the body using ice packs, cooling blankets, or other cooling methods to slow down metabolic processes and minimize cellular degradation. The goal is to halt the deterioration that would typically occur after legal death and prepare the body for cryopreservation.

Once the body is stabilized, the next crucial step is perfusion, which involves replacing the blood and other bodily fluids with cryoprotective agents. Cryoprotective agents are specially formulated solutions that protect the cells during freezing and prevent ice crystal formation. The process of perfusion typically begins by accessing the major blood vessels, such as the carotid artery and jugular vein, in whole-body cryonics, or the femoral artery and vein, in brain-only preservation.

During perfusion, the cryoprotective agents are gradually introduced into the circulatory system while simultaneously removing blood and other fluids from the body. This process allows the cryoprotective agents to penetrate the cells and replace the water molecules, reducing the formation of ice crystals during freezing. Cryoprotective agents commonly used in cryonics include glycerol, dimethyl sulfoxide (DMSO), and ethylene glycol.

After perfusion, the cooling process begins. The body or brain is cooled gradually to a temperature below freezing, typically around -130 degrees Celsius (-202 degrees Fahrenheit). Controlled cooling is essential to minimize the formation of ice crystals, which can cause damage to cellular structures. Cooling is achieved using specialized cooling devices or chambers that allow for precise temperature control.

Once the desired temperature is reached, the final stage of cryonics procedures commences: vitrification and long-term storage. Vitrification involves further cooling the body or brain to temperatures below -196 degrees Celsius (-321 degrees Fahrenheit), at which point the tissues undergo a transition from a liquid to a solid state, without the formation of ice crystals. This solidified state is similar to glass and is designed to preserve cellular structures and minimize damage.

To achieve long-term storage, the cryonically preserved individual is placed in a specially designed storage container filled with liquid nitrogen. Liquid nitrogen maintains a stable temperature of -196 degrees Celsius (-321 degrees Fahrenheit), effectively halting all metabolic processes and slowing down the degradation of tissues. The storage containers are designed to maintain the integrity of the cryopreserved individual and to minimize temperature fluctuations and potential contamination.

Regular maintenance and monitoring of the storage facilities are critical to ensuring the integrity of the cryopreserved samples. Cryonics organizations have strict protocols in place, including regular checks of liquid nitrogen levels, temperature monitoring, and security measures to prevent unauthorized access.

It is important to note that cryonics procedures are continuously evolving, with ongoing research and advancements in the field. Cryonics organizations are continually refining their techniques, improving cryoprotective solutions, and exploring innovative technologies to enhance the preservation and potential future revival of cryonically preserved individuals.

## Standby and Stabilization

Standby and stabilization are crucial components of the cryonics process, playing a vital role in ensuring the preservation of the human body or brain following legal death. These initial stages involve prompt response, rapid cooling, and the implementation of measures to minimize cellular degradation. Below we will explore the processes of standby and stabilization in relation to cryonics.

Standby refers to the readiness of cryonics organizations to respond immediately after legal death occurs. Cryonics organizations maintain standby teams that are available around the clock to be notified promptly when legal death is pronounced. These teams are equipped with the necessary tools, equipment, and expertise to initiate the cryonics procedures as quickly as possible.

When cryonics organizations are notified of legal death, the standby team mobilizes to the location of the deceased individual. The goal is to minimize the time elapsed between legal death and the initiation of cryonics procedures, as this time is critical for preserving the integrity of the body or brain.

Stabilization is the first step in cryonics after the arrival of the standby team. The primary objective of stabilization is to slow down metabolic processes and minimize cellular degradation until the cryonics procedures can be implemented.

The process of stabilization typically begins by cooling the body or brain using cooling blankets, ice packs, or other cooling methods. This rapid cooling serves to slow down the chemical reactions in the body and reduce the rate of cellular decay. Cooling also helps protect the brain, which is particularly vulnerable to damage due to its high metabolic activity.

In addition to cooling, other measures may be taken during the stabilization phase. Administration of medications, such as anticoagulants or antioxidants, may be employed to mitigate potential damage caused by blood clots or oxidative stress. The application of cardiopulmonary support, including chest compressions or mechanical ventilation, may be initiated to maintain oxygenation and blood flow to vital organs.

The goal of stabilization is to create a favorable environment for the subsequent cryonics procedures. By cooling the body and administering appropriate medications, the cryonics team aims to delay cellular degradation and preserve the structure of tissues, enabling the subsequent steps of cryopreservation to be carried out more effectively.

The time frame for stabilization is typically short, as cryonics organizations strive to begin the cryopreservation process as quickly as possible. The precise duration of the stabilization phase can vary depending on factors such as the distance to the location, response time, and the condition of the deceased individual.

Efficient communication and coordination between cryonics organizations and medical professionals are essential during standby and stabilization. Cryonics organizations work closely with medical personnel, ensuring that the necessary permissions and legal requirements are met. Collaboration between cryonics organizations and medical professionals facilitates a smooth transition from legal death to the cryonics procedures, enhancing the likelihood of successful preservation.

It is important to note that standby and stabilization are critical but temporary measures. The ultimate aim is to initiate the subsequent steps of cryopreservation, including perfusion, cooling, vitrification, and long-term storage. These later stages of the cryonics process are fundamental to achieving the preservation of cellular structures and maximizing the potential for future revival.

## Cooling and Transportation

Cooling and transportation are crucial steps in the cryonics process, ensuring that the body or brain is preserved at low temperatures and transported to the cryonics facility for long-term storage. These stages play a vital role in maintaining the structural integrity of the tissues and minimizing the risk of cellular damage. Below we will explore the processes of cooling and transportation in relation to cryonics.

After the stabilization phase, the body or brain is prepared for cooling. The primary objective of cooling is to lower the temperature of the tissues to cryogenic levels, typically around -196 degrees Celsius (-321 degrees Fahrenheit), at which biological processes virtually come to a halt. Cooling is critical for minimizing further cellular damage and preserving the structures necessary for potential future revival.

To achieve the necessary low temperatures, cryonics organizations employ various cooling techniques. One commonly used method is the use of dry ice or liquid nitrogen. Dry ice, which is frozen carbon dioxide, is often used for short-term cooling and transportation purposes. Liquid nitrogen, on the other hand, provides a colder and more stable environment for long-term storage.

The body or brain is carefully placed in a cooling container, such as a specially designed dewar, which is capable of maintaining extremely low temperatures. The container is then filled with dry ice or liquid nitrogen to create the necessary cryogenic environment. The cooling process must be conducted gradually to prevent thermal shock, which could potentially damage the tissues.

During transportation, cryonics organizations take great care to ensure the preservation of the cooled body or brain. The containers used for transportation are specifically designed to provide insulation and maintain the low temperatures required for cryopreservation. The containers are often well-insulated and equipped with monitoring systems to ensure temperature stability throughout the journey.

Transportation of cryonics patients can be challenging, as it often involves long distances and logistical considerations. Cryonics organizations work closely with transportation providers to ensure the safe and efficient transfer of the cryopreserved individuals. The containers are typically sealed and secured to prevent any potential damage during transit.

In addition to temperature control and physical security, cryonics organizations must also comply with legal and regulatory requirements during transportation. These requirements vary depending on the jurisdiction and may involve obtaining permits or adhering to specific transport protocols. Cryonics organizations collaborate closely with relevant authorities to ensure compliance and a smooth transportation process.

The success of the cooling and transportation stages greatly depends on the efficiency of the procedures and the expertise of the cryonics organizations involved. Proper training and adherence to protocols are essential to maintaining the quality of the cryopreserved tissues during transportation. Regular monitoring and quality control measures are implemented to ensure that the low temperatures are maintained and any potential issues are promptly addressed.

It is worth noting that the cooling and transportation stages are just one part of the cryonics process. They are the intermediate steps between stabilization and the final stage of long-term storage. Once the cryopreserved individuals reach the cryonics facility, they undergo further procedures, including vitrification and storage in specialized facilities designed to maintain ultra-low temperatures.

# Perfusion and Vitrification

Perfusion and vitrification are two critical steps in the cryonics process, contributing to the successful preservation of tissues and organs for potential future revival. These stages involve the removal of blood and other fluids from the body or brain, followed by the introduction of cryoprotectants and the transition to a glass-like state. Below we will delve into the processes of perfusion and vitrification in relation to cryonics.

Perfusion is the process of replacing the blood and other fluids in the body with a cryoprotective solution. Cryoprotectants are substances that help prevent ice formation and cellular damage during the freezing process. They act by reducing the freezing point of the tissues and protecting cell structures from the harmful effects of ice crystals. The primary goal of perfusion is to distribute the cryoprotectant throughout the body or brain, replacing the fluids and ensuring uniform protection.

To initiate perfusion, cryonics organizations typically use a combination of anticoagulants and cryoprotective solutions. Anticoagulants prevent blood clotting and enable a smooth flow of the perfusate through the blood vessels. Cryoprotective solutions, such as glycerol or dimethyl sulfoxide (DMSO), are introduced gradually to replace the blood and interstitial fluids.

Perfusion requires careful coordination and expertise to ensure proper distribution of the cryoprotective solution. Cryonics professionals use specialized equipment, such as pumps and cannulas, to control the flow and ensure the cryoprotectant reaches all parts of the body or brain. The process is often conducted at low temperatures to minimize metabolic activity and reduce the risk of tissue damage.

Once perfusion is complete, the next step is vitrification. Vitrification involves cooling the tissues to extremely low temperatures, typically below -120 degrees Celsius (-184 degrees Fahrenheit), at which the cryoprotectants solidify without forming ice crystals. This process is crucial for avoiding ice formation, which can lead to cellular damage and compromise the structural integrity of the tissues.

During vitrification, the cryoprotected tissues are gradually cooled while maintaining a high concentration of cryoprotectants. The cryoprotectants undergo a glass transition, transforming into an amorphous solid that preserves the cellular structures. The key principle of vitrification is to prevent the formation of ice crystals by avoiding rapid cooling and maintaining high cryoprotectant concentrations.

Achieving successful vitrification requires precise control of cooling rates and cryoprotectant concentrations. Cryonics organizations utilize sophisticated cooling protocols and equipment to ensure a gradual and controlled cooling process. Various monitoring techniques, such as temperature sensors and imaging technologies, are employed to monitor the vitrification process and ensure the preservation of tissues.

The advancements in vitrification techniques have played a significant role in improving the prospects of successful cryopreservation. By transitioning to a glass-like state, the tissues can be stored at ultra-low temperatures without the damaging effects of ice formation. However, it is important to note that the long-term stability and reversibility of vitrification remain subjects of ongoing research and development in the field of cryonics.

Perfusion and vitrification are complex processes that require careful planning, expertise, and technological advancements. Cryonics organizations continuously strive to improve these techniques and optimize the preservation methods to enhance the chances of successful cryopreservation. Ongoing research and collaboration within the cryonics community are focused on refining the protocols, identifying new cryoprotectants, and exploring innovative approaches to achieve better outcomes.

# The Cryonics Industry: Key Players and Facilities

The cryonics industry comprises a network of organizations, facilities, and professionals dedicated to the preservation and potential revival of human bodies and brains using cryopreservation techniques. Below we will explore the key players and facilities in the cryonics industry and their contributions to advancing the field.

The Cryonics Institute (CI) is one of the most prominent organizations in the cryonics industry. Founded in 1976, CI is a nonprofit organization located in Clinton Township, Michigan, USA. It provides cryopreservation services to its members and operates a state-of-the-art cryonics facility. CI offers both whole-body and neuro (brain-only) preservation options, utilizing vitrification as the primary method of cryopreservation.

Another leading organization in the cryonics industry is Alcor Life Extension Foundation. Established in 1972 and headquartered in Scottsdale, Arizona, USA, Alcor is a nonprofit organization that specializes in cryopreservation services. Alcor is known for its research and development efforts in cryonics technology and has contributed significantly to advancements in perfusion and vitrification techniques. The organization offers whole-body and neuro cryopreservation options and operates a dedicated cryonics facility.

KrioRus is the first cryonics organization in Russia and Eurasia. Founded in 2005, KrioRus is a commercial entity that provides cryonics services, including both whole-body and neuro preservation. The organization operates a facility in Moscow and has established collaborations with international cryonics organizations to ensure the availability of cryopreservation services to its members.

In addition to these major players, there are several other organizations and facilities around the world that contribute to the cryonics industry. Examples include the Oregon Cryonics facility in the United States, which offers cryopreservation services and focuses on research and development in cryonics technology. Cryonics UK is another notable organization providing cryopreservation services to its members in the United Kingdom.

Cryonics facilities are specifically designed to accommodate the cryopreservation process. These facilities feature specialized equipment, such as perfusion systems, cooling chambers, and long-term storage units. They adhere to stringent protocols to ensure the preservation and maintenance of the cryopreserved bodies and brains.

Long-term storage is a critical aspect of cryonics facilities. Organizations typically store their patients' remains in large cryostorage containers known as dewars, which are filled with liquid nitrogen to maintain temperatures below -196 degrees Celsius (-321 degrees Fahrenheit). These facilities employ rigorous monitoring systems to ensure the stability and safety of the cryopreserved individuals.

It is important to note that the cryonics industry operates within legal and regulatory frameworks specific to each jurisdiction. The industry faces unique challenges due to the novel nature of its practices and the ethical considerations associated with the preservation and potential revival of human life. Organizations in the field work closely with legal and regulatory authorities to ensure compliance and promote the development of appropriate regulations.

The key players and facilities in the cryonics industry play a vital role in advancing the field through research, development, and the provision of cryopreservation services. Their efforts contribute to improving cryonics techniques, enhancing storage methods, and expanding access to cryopreservation for individuals who seek this option for their end-of-life arrangements.

## Major Cryonics Organizations and Their Histories

Cryonics, the practice of preserving human bodies or brains at extremely low temperatures with the aim of future revival, has gained attention and interest in recent decades. Below we will explore the major cryonics organizations and their histories, highlighting their contributions to the field and the advancements they have made in cryonics technology.

The Cryonics Institute (CI) is one of the oldest and most well-known cryonics organizations. It was founded in 1976 by Robert Ettinger, often referred to as the "father of cryonics." Ettinger's book "The Prospect of Immortality" laid the foundation for the concept of cryonics and inspired the establishment of CI. Located in Clinton Township, Michigan, CI has been at the forefront of cryonics research and development. It offers both whole-body and neuro (brain-only) preservation options using vitrification as the primary method of cryopreservation.

Another prominent cryonics organization is the Alcor Life Extension Foundation, which was founded in 1972 by Fred and Linda Chamberlain. Alcor is based in Scottsdale, Arizona, and has played a significant role in advancing cryonics technology. The organization has conducted extensive research in cryopreservation techniques, including the development of advanced perfusion and vitrification methods. Alcor offers both whole-body and neuro cryopreservation options and has a dedicated research facility in addition to its cryonics facility.

KrioRus is the first and largest cryonics organization in Russia and Eurasia. Founded in 2005 by Valeriya and Danila Medvedev, KrioRus aims to provide cryopreservation services to individuals in the region. The organization has been instrumental in expanding access to cryonics for individuals who may not have had the opportunity otherwise. KrioRus operates a facility in Moscow and has established collaborations with international cryonics organizations to ensure the availability of cryopreservation services to its members.

The Cryonics Institute, Alcor, and KrioRus are just a few examples of the major cryonics organizations. There are several other organizations worldwide that offer cryopreservation services, including organizations such as Oregon Cryonics in the United States and Cryonics UK in the United Kingdom.

Each of these organizations has made significant contributions to the field of cryonics through research, technological advancements, and the provision of cryopreservation services. They have invested resources into improving the cryopreservation process, developing better cryoprotectants, and enhancing the long-term storage methods to ensure the preservation of the patients' remains.

Over the years, cryonics organizations have faced various challenges, including legal and ethical considerations. The practice of cryonics raises questions about the definition of death, the potential for future revival, and the societal implications of such technology. Cryonics organizations work closely with legal and regulatory authorities to ensure compliance with existing regulations and to advocate for the development of appropriate regulations that address the unique aspects of cryonics.

In addition to providing cryopreservation services, these organizations engage in public outreach and education to increase awareness and understanding of cryonics. They organize conferences, publish scientific papers, and collaborate with researchers and professionals in related fields to share knowledge and advance the science of cryonics.

## A Look Inside Cryonics Facilities

Cryonics facilities are specialized facilities designed to accommodate the unique requirements of cryopreserving human bodies or brains. These facilities play a crucial role in the cryonics process, providing the necessary infrastructure and equipment to ensure the preservation and long-term storage of cryopreserved individuals. Below we will take a look inside cryonics facilities and explore their key features and functions.

Cryonics facilities are purpose-built or modified facilities specifically designed to meet the needs of cryopreservation. These facilities adhere to strict protocols and standards to ensure the integrity and stability of the cryopreserved remains. They are equipped with state-of-the-art technology and infrastructure to support the cryonics process.

One essential feature of a cryonics facility is the cryostorage area. This area consists of large containers called dewars, which are designed to hold cryopreserved individuals. Dewars are typically filled with liquid nitrogen to maintain extremely low temperatures, usually below -196 degrees Celsius (-321 degrees Fahrenheit). The cryostorage area is carefully monitored and maintained to ensure the stability and safety of the cryopreserved individuals.

Cryonics facilities also house specialized equipment necessary for the cryopreservation process. This includes perfusion systems, cooling chambers, and monitoring devices. Perfusion systems are used to replace the body's fluids with cryoprotective solutions, ensuring uniform distribution of cryoprotectants throughout the tissues. Cooling chambers are used to gradually cool the cryopreserved remains, often employing controlled cooling rates to prevent ice formation. Monitoring devices, such as temperature sensors and imaging technologies, are employed to continuously monitor the status of the cryopreserved individuals and ensure the preservation conditions are optimal.

To maintain the cryopreserved individuals, cryonics facilities implement rigorous protocols for long-term storage. This involves regular monitoring of temperature and liquid nitrogen levels, as well as periodic maintenance and inspections of the cryostorage area and equipment. Facilities may have backup systems and contingency plans in place to address any unforeseen circumstances or technical issues that may arise.

Another important aspect of cryonics facilities is their commitment to security and confidentiality. The confidentiality of the cryopreserved individuals' identities and personal information is of utmost importance. Cryonics organizations maintain strict privacy policies to protect the privacy and confidentiality of their members.

Cryonics facilities also serve as hubs for research and development in the field of cryonics. Many organizations conduct ongoing research to improve cryopreservation techniques, develop new cryoprotectants, and explore innovative approaches to enhance the prospects of future revival. Facilities may have dedicated research areas where scientists and researchers work on advancing the field of cryonics.

In addition to their technical and scientific functions, cryonics facilities often provide educational and outreach opportunities to promote understanding and awareness of cryonics. This may include hosting public events, offering tours, and collaborating with educational institutions to disseminate information about the science and philosophy of cryonics.

Legal and regulatory compliance is another crucial aspect of cryonics facilities. Organizations and facilities must adhere to the laws and regulations specific to their jurisdiction. They work closely with legal and regulatory authorities to ensure compliance, promote best practices, and advocate for the development of appropriate regulations that address the unique aspects of cryonics.

# The Evolution of Cryonic Storage Technologies

The evolution of cryonic storage technologies has been a critical factor in the advancement of the cryonics field. These technologies are responsible for ensuring the long-term preservation and stability of cryopreserved individuals. Over the years, significant progress has been made in developing more effective and efficient storage methods. Below we will explore the evolution of cryonic storage technologies and their contributions to the field of cryonics.

Early cryonic storage methods relied on dewars filled with liquid nitrogen to maintain low temperatures for long-term storage. Liquid nitrogen provides a stable and extremely cold environment, but it presents challenges in terms of temperature control and maintenance. As the cryonics field progressed, researchers and engineers sought to develop more sophisticated and reliable storage technologies.

One notable advancement in cryonic storage technologies is the introduction of dry nitrogen vapor storage. This technique involves exposing the cryopreserved individuals to a controlled flow of dry nitrogen vapor rather than immersing them in liquid nitrogen. Dry nitrogen vapor storage offers several advantages, including improved temperature control, reduced risk of contamination, and easier access to the samples for research or potential future interventions.

Another significant development in cryonic storage is the use of insulated pods or capsules for individual cryopreserved individuals. These pods, also known as cryocapsules, are designed to provide enhanced insulation and protection for the cryopreserved remains. They are often made of advanced materials with low thermal conductivity to minimize heat transfer and maintain stable temperatures. Cryocapsules can be stored within larger dewars or cryostorage containers to provide additional layers of insulation and security.

Advancements in material science have also contributed to the evolution of cryonic storage technologies. The development of advanced insulating materials, such as multilayered insulation (MLI), has improved the efficiency of cryostorage systems. MLI consists of layers of reflective material separated by thin spacers, which reduce heat transfer by reflecting thermal radiation. These insulating materials help to maintain stable temperatures and minimize the consumption of cryogenic fluids.

In recent years, cryonics organizations have explored the use of advanced cryogenic storage systems, such as liquid nitrogen vapor phase storage. This storage method involves storing cryopreserved individuals in an environment of nitrogen vapor above the liquid nitrogen. By utilizing the vapor phase, organizations can reduce the risk of liquid nitrogen exposure and improve temperature control. Liquid nitrogen vapor phase storage systems offer enhanced safety features and more efficient use of cryogenic fluids.

The introduction of automated storage and retrieval systems has also revolutionized cryonic storage. These systems utilize robotic mechanisms to retrieve and transport cryocapsules within the storage facility. Automated storage systems enable efficient organization and retrieval of cryopreserved individuals, reduce the risk of manual handling errors, and optimize the use of storage space.

Advancements in monitoring and control systems have further enhanced cryonic storage technologies. Modern cryonic storage facilities utilize sophisticated sensors, data logging systems, and remote monitoring capabilities to ensure stable storage conditions. These systems continuously monitor parameters such as temperature, pressure, and liquid nitrogen levels, providing real-time information and alerts for timely intervention if needed.

It is important to note that the development of cryonic storage technologies is an ongoing process. Cryonics organizations and researchers continuously explore new materials, designs, and strategies to improve the efficiency and effectiveness of cryonic storage. They collaborate with experts in various fields, including engineering, material science, and cryogenics, to push the boundaries of what is possible in cryonic storage.

# Cryonics and Medicine: Current Research and Future Possibilities

Cryonics, the practice of preserving human bodies or brains at ultra-low temperatures with the goal of future revival, intersects with various areas of medicine and has the potential to impact medical research and advancements. Below we will explore the current research efforts and future possibilities of cryonics in relation to the field of medicine.

One area where cryonics intersects with medicine is organ transplantation. Cryopreservation techniques used in cryonics could potentially extend the preservation time of organs and tissues, allowing for more efficient transplantation procedures. The ability to preserve organs for longer periods could increase the availability of viable organs for transplantation, potentially saving more lives and reducing organ shortage.

Researchers are exploring the application of cryopreservation techniques in the field of regenerative medicine. Cryopreserved tissues and cells can be used as a valuable resource for developing therapies and studying diseases. Cryonics may facilitate the creation of tissue banks, where cryopreserved cells and tissues can be stored for future use in research or transplantation purposes.

Another area of interest is the preservation of neural tissue for studying brain disorders and neurological conditions. Cryonics offers the potential to retain the structural and molecular integrity of the brain, allowing researchers to investigate the mechanisms underlying various neurological disorders. This could lead to a deeper understanding of these conditions and the development of novel treatments.

Cryopreservation techniques also have implications for cryopreserving patients with terminal illnesses or conditions for whom current medical treatments are inadequate. It is conceivable that advancements in medical technologies and therapies in the future could enable the revival and treatment of these patients. Cryonics provides a potential bridge between the present and the future, offering the possibility of accessing future medical interventions and advancements.

Current research in cryonics is focused on improving the cryopreservation process and exploring new approaches to enhance the chances of successful revival. Scientists are investigating the development of better cryoprotectants, which are substances used to protect tissues during the freezing process. The goal is to find cryoprotectants that are less toxic, more effective, and capable of preserving cellular structures and functions.

Advancements in nanotechnology may also play a role in the future of cryonics and medicine. Nanotechnology could potentially be used to repair cellular damage caused by the cryopreservation process or to enhance the revival and restoration of cryopreserved tissues. Nanoscale technologies might enable precise control over cellular and molecular processes, aiding in the revival and rejuvenation of cryopreserved individuals.

One of the challenges faced in cryonics research is the development of reliable methods for rewarming and revival. The process of reversing the cryopreservation and restoring cellular functions without causing further damage is a complex task that requires significant advancements in medical and technological fields. Scientists are exploring various strategies, such as nanowarming and molecular repair techniques, to overcome these challenges.

Despite the ongoing research and advancements, it is important to acknowledge the limitations and ethical considerations surrounding cryonics. The scientific community continues to debate the plausibility and feasibility of future revival, and the ethical implications of cryonics are subject to various perspectives and opinions.

## Advances in Regenerative Medicine

Regenerative medicine is an exciting and rapidly advancing field that holds great promise for the future of healthcare. It focuses on harnessing the body's natural healing processes to restore or replace damaged tissues and organs. In recent years, significant advances have been made in regenerative medicine, and these developments intersect with the concept of cryonics, offering potential synergies and implications for the field. Below we will explore the key advances in regenerative medicine and their relationship to cryonics.

One of the most notable advancements in regenerative medicine is the development of stem cell therapies. Stem cells have the remarkable ability to differentiate into various cell types and have the potential to regenerate damaged tissues. Researchers have made significant progress in isolating and manipulating different types of stem cells, including embryonic stem cells, induced pluripotent stem cells, and adult stem cells. These cells can be used to replace or repair damaged cells and tissues in a wide range of medical conditions.

Cryonics, on the other hand, aims to preserve the body or brain at ultra-low temperatures for future revival. The preservation of stem cells through cryopreservation techniques has immense potential in regenerative medicine. Cryopreserved stem cells can be stored and used at a later time, providing a valuable resource for tissue regeneration and repair. Stem cells preserved through cryonics could be utilized in personalized regenerative medicine, where a person's own cells are used to regenerate tissues and organs, reducing the risk of rejection.

Advancements in tissue engineering and 3D bioprinting have also revolutionized regenerative medicine. Tissue engineering involves creating functional tissues in the laboratory by combining cells, biomaterials, and growth factors. 3D bioprinting takes this a step further by precisely depositing layers of cells and biomaterials to create complex three-dimensional structures. These techniques have been used to create artificial organs, such as skin, blood vessels, and even small-scale organs like the liver and heart. Cryonics may provide a means to preserve these engineered tissues, allowing for future transplantation and replacement of damaged organs.

In the field of regenerative medicine, researchers are also exploring the potential of gene therapy. Gene therapy involves modifying or replacing genes to treat or prevent diseases. This approach holds great promise for addressing genetic disorders and inherited conditions. Cryonics could play a role in preserving genetic material, such as DNA, for future gene therapy applications. Preserving genetic information through cryopreservation may allow for the modification and correction of genes in the future, offering potential treatments for a wide range of genetic conditions.

Another significant advancement in regenerative medicine is the use of biomaterials and scaffolds to support tissue regeneration. These materials provide a framework for cells to grow and organize themselves into functional tissues. Cryonics can preserve biomaterials and scaffolds, ensuring their availability for future use in tissue engineering and regeneration. Cryopreservation techniques may help maintain the structural integrity and functionality of these materials, enhancing their effectiveness in regenerative medicine applications.

Furthermore, the field of regenerative medicine is exploring the potential of exosomes and other extracellular vesicles in tissue regeneration. Exosomes are tiny vesicles released by cells that contain various bioactive molecules, such as proteins, nucleic acids, and growth factors. These vesicles play a crucial role in cell-to-cell communication and can influence the behavior of recipient cells. Cryonics can preserve exosomes, allowing for their storage and future utilization in regenerative therapies. Preserving the exosome cargo through cryopreservation ensures the availability of these powerful signaling molecules for future tissue regeneration strategies.

It is important to note that while regenerative medicine has made remarkable progress, there are still challenges and ethical considerations to address. The clinical translation of regenerative therapies requires rigorous testing, ensuring their safety and efficacy.

# The Potential of Nanotechnology

Nanotechnology, the manipulation of matter on an atomic and molecular scale, has emerged as a transformative field with vast potential in various scientific disciplines. In recent years, nanotechnology has garnered attention in the realm of cryonics, offering exciting possibilities for improving preservation techniques and enhancing the chances of successful future revival. Below we will explore the potential of nanotechnology in the context of cryonics.

One of the primary challenges in cryonics is the preservation of the delicate cellular structures within the body or brain. Traditional cryopreservation methods, such as vitrification, aim to prevent ice formation and minimize damage during the cooling process. However, even with advanced techniques, some cellular structures and components can still be affected, leading to potential damage upon revival. This is where nanotechnology comes into play.

Nanotechnology offers the ability to manipulate matter at the nanoscale, allowing precise control over the properties and behavior of materials. Nanoparticles, nanomaterials, and nanodevices can be designed and engineered to interact with biological systems in unique ways. In the context of cryonics, nanotechnology can play a crucial role in preserving and protecting cellular structures during the cryopreservation process.

One area where nanotechnology can contribute to cryonics is in the development of cryoprotective agents (CPAs). CPAs are chemicals used to reduce ice formation and prevent damage to cells during the freezing process. Current CPAs have limitations, such as toxicity and limited penetration into tissues. Nanoparticles can be used to encapsulate CPAs, improving their delivery and enhancing their protective effects. These nanoparticles can be engineered to release CPAs at specific temperatures or in response to certain stimuli, ensuring optimal protection during cryopreservation.

Furthermore, nanotechnology can facilitate the monitoring and control of cryopreservation conditions. Nanosensors embedded within the cryopreservation medium can provide real-time information on temperature, pressure, and other critical parameters. This enables precise monitoring and adjustment of cryopreservation conditions, ensuring optimal preservation and minimizing the risk of damage. Nanotechnology can also aid in the development of smart cryopreservation systems that automatically adapt to changing conditions, further improving the preservation process.

Another exciting application of nanotechnology in cryonics is the development of nanorobots or nanomachines. These tiny devices, typically ranging from a few nanometers to a few micrometers in size, can navigate through the body and perform specific tasks at the cellular or molecular level. In the context of cryonics, nanorobots can be designed to repair cellular damage, remove toxins, or facilitate the regeneration of tissues upon revival. These nanorobots could be programmed to target specific areas of damage and deliver therapeutic agents, enhancing the chances of successful revival and functional recovery.

Nanotechnology also holds promise in the field of nanomedicine, which involves the use of nanoscale tools and techniques for medical applications. Nanoparticles can be functionalized with specific molecules, such as antibodies or targeting ligands, to selectively deliver therapeutic agents to damaged or diseased cells. In the context of cryonics, nanomedicine approaches can be employed to repair cellular damage, regenerate tissues, or stimulate the body's natural healing processes upon revival. This could significantly enhance the viability and functional recovery of cryopreserved individuals.

Moreover, nanotechnology has the potential to revolutionize imaging techniques used in cryonics. Current imaging methods, such as electron microscopy, have limitations in terms of resolution and depth penetration. Nanoscale imaging techniques, such as scanning probe microscopy or super-resolution microscopy, can provide unprecedented details of cellular structures and molecular interactions. These advanced imaging techniques can aid in the assessment of preservation quality, identification of cellular damage, and evaluation of tissue regeneration upon revival.

# Neurological Preservation and Recovery

Neurological preservation and recovery are critical aspects of cryonics, as the brain is the seat of consciousness and holds our memories, thoughts, and identities. Cryonics seeks to preserve the brain at ultra-low temperatures in the hope of future revival and restoration of neurological function. Below we will explore the challenges and possibilities of neurological preservation and recovery in the context of cryonics.

Preserving the delicate neural structures and connections within the brain is a complex task. The traditional approach to cryonics involves vitrification, a process that replaces water with cryoprotective agents to prevent ice formation and minimize damage. While vitrification has shown promise, there are still concerns about the preservation of fine neuronal structures and synapses. The challenge lies in maintaining the integrity of neural circuits, which are essential for the restoration of brain function.

Advances in neuroscience and neurobiology have shed light on the complexity of the brain and its intricate networks. Understanding the organization and connectivity of the brain is crucial for successful neurological preservation and recovery. Mapping techniques, such as connectomics, aim to comprehensively chart the connections between neurons and decipher the neural circuits responsible for various functions. This knowledge can guide the preservation process and help ensure that critical neural connections are preserved during cryopreservation.

In recent years, researchers have also made significant progress in the field of brain-machine interfaces (BMIs) or neural prosthetics. BMIs involve the direct connection between the brain and external devices, enabling individuals to control prosthetic limbs or communicate using their thoughts. This technology holds promise for neurological recovery in cryonics. If a cryopreserved individual can be successfully revived, BMIs could potentially bridge the gap between the restored brain and the external world, facilitating communication and mobility.

Another area of research relevant to neurological preservation and recovery is the field of neuroregeneration. Neuroregeneration aims to stimulate the regrowth of damaged neurons and restore neural connections. Scientists are exploring various strategies, including stem cell therapy, growth factor administration, and gene therapy, to encourage the regrowth and repair of neurons. If successful, these approaches could potentially be applied to revive cryopreserved individuals and restore neural function.

Furthermore, advancements in nanotechnology and nanomedicine offer exciting possibilities for neurological recovery. Nanoparticles and nanodevices can be designed to interact with neurons and deliver therapeutic agents directly to damaged areas of the brain. These nanoscale tools can provide targeted treatment and promote the regeneration of neural tissues. Additionally, nanosensors can be used to monitor brain activity and provide real-time feedback during the revival process, assisting in the evaluation and adjustment of neurological recovery strategies.

Ethical considerations also come into play when discussing neurological preservation and recovery in cryonics. Questions arise regarding personal identity, consciousness, and the restoration of a person's subjective experience. While cryonics seeks to preserve the physical substrate of the brain, it is unclear how individual memories and consciousness would be restored in a revived state. These philosophical and ethical dilemmas require careful consideration as the field of cryonics advances.

# The Cost of Cryonics: Financial Planning and Funding Options

The cost of cryonics is a significant consideration for individuals who are interested in pursuing this unique form of life preservation. Cryonics involves the freezing and storage of a person's body or brain in the hope of future revival and medical advancements. Below we will explore the financial planning and funding options associated with cryonics.

The cost of cryonics can vary depending on several factors, including the chosen cryonics organization, the type of preservation (whole-body or neuro), and additional services such as standby, transportation, and long-term storage. It is important to note that cryonics is a long-term commitment, and costs may be spread out over many years.

One of the primary expenses in cryonics is the initial cryopreservation procedure. This typically involves the use of cryoprotectants to replace water in the body or brain with substances that can prevent ice formation and preserve cellular structures. The cost of this procedure can range from tens of thousands to hundreds of thousands of dollars, depending on the organization and the level of service provided.

In addition to the cryopreservation procedure, there are ongoing costs associated with cryonics. These include the cost of long-term storage, which involves the maintenance and monitoring of the cryopreserved remains over an extended period. The expenses for storage can be significant, and individuals are often required to make arrangements for funding this aspect of cryonics.

Given the substantial costs involved, it is essential for individuals interested in cryonics to engage in careful financial planning. Here are some strategies and funding options to consider:

1. Life Insurance: Many individuals opt to purchase a life insurance policy to cover the expenses of cryonics. By naming the cryonics organization or a trust as the beneficiary, the policy's payout can be specifically designated for cryonics-related costs.

2. Trusts and Estate Planning: Establishing a trust or including cryonics-related provisions in an estate plan can help ensure that funds are set aside for cryonics expenses. Working with an attorney experienced in estate planning can help individuals structure their financial arrangements appropriately.

3. Prepaid Contracts: Some cryonics organizations offer the option to prepay for cryopreservation services. This allows individuals to lock in current pricing and make arrangements for the future. However, it is crucial to research and understand the terms and conditions of such contracts before committing.

4. Crowdfunding: In recent years, crowdfunding platforms have emerged as a popular way for individuals to raise funds for cryonics. Through online campaigns, individuals can share their stories and seek financial support from friends, family, and the broader community.

5. Saving and Investing: Another approach to funding cryonics is to save and invest money over time. By setting aside funds in a dedicated cryonics fund or investment account, individuals can gradually accumulate the necessary funds to cover cryonics expenses.

It is important to note that financial planning for cryonics should be done in consultation with financial advisors and legal professionals. They can provide guidance on the most appropriate strategies for individual circumstances and help ensure that funds are properly allocated and protected.

# Understanding the Expenses Involved

Cryonics, the practice of preserving the human body or brain at extremely low temperatures, has gained attention as a potential means to extend life or facilitate future medical advancements. While cryonics offers hope for the future, it is important to understand the expenses involved in this unique undertaking. Below we will explore the various expenses associated with cryonics, shedding light on the financial aspects of this field.

One of the primary expenses in cryonics is the cryopreservation process itself. Cryopreservation involves the use of cryoprotectants to replace water in the body or brain with substances that can prevent ice formation and preserve cellular structures. This complex procedure requires specialized equipment and highly trained personnel. The cost of cryopreservation can range from tens of thousands to hundreds of thousands of dollars, depending on the cryonics organization and the level of service provided.

In addition to the initial cryopreservation, there are ongoing expenses associated with cryonics. These include the costs of long-term storage and maintenance. Cryonics organizations maintain specialized facilities where the cryopreserved remains are stored and monitored over extended periods. The expenses for storage can be substantial due to the need for advanced technology, constant monitoring, and dedicated personnel.

Furthermore, it is essential to consider the costs associated with standby and transportation. Standby refers to the process of having a team ready to respond and begin cryopreservation procedures immediately upon legal death. This service ensures timely intervention to minimize any potential damage to the body or brain. Transportation costs are incurred when the cryopreserved remains need to be transported to the cryonics facility. These expenses depend on factors such as the distance traveled and the mode of transportation.

To fund cryonics expenses, individuals have several options:

1. Prepayment: Some cryonics organizations offer the option to prepay for cryopreservation services. Prepayment allows individuals to lock in current pricing and make financial arrangements in advance. It is important to thoroughly research and understand the terms and conditions associated with prepayment options.

2. Life Insurance: Many individuals opt to purchase life insurance policies specifically designated to cover cryonics expenses. By naming the cryonics organization or a trust as the beneficiary, the insurance payout can be used to cover the costs associated with cryopreservation and storage.

3. Trusts and Estate Planning: Establishing a trust or including cryonics provisions in an estate plan can ensure that funds are allocated for cryonics expenses. This approach requires working closely with legal professionals who specialize in estate planning to structure the financial arrangements appropriately.

4. Crowdfunding: In recent years, crowdfunding platforms have become popular for raising funds for various purposes, including cryonics. Individuals can share their stories and seek financial support from friends, family, and the wider community.

5. Saving and Investment: Another approach is to save and invest funds over time to cover cryonics expenses. By setting aside money in a dedicated cryonics fund or investment account, individuals can gradually accumulate the necessary funds.

It is important to note that the cost of cryonics is influenced by several factors, including the cryonics organization chosen, the level of service desired, and the additional services required. It is crucial to thoroughly research different cryonics organizations, understand their services and pricing structures, and engage in comprehensive financial planning to ensure that the necessary funds are available.

# Life Insurance and Cryonics Trusts

Cryonics, the practice of preserving the human body or brain at extremely low temperatures, offers the possibility of future revival and medical advancements. While cryonics provides hope for the future, it is crucial to consider the financial aspects of this unique endeavor. Below we will explore two funding options for cryonics: life insurance and cryonics trusts. These financial tools can help individuals ensure that the necessary funds are available to cover cryonics expenses.

Life Insurance:

Life insurance is a common financial instrument used to provide financial protection for loved ones after one's death. In the context of cryonics, life insurance can be utilized to fund cryopreservation and related expenses. Here's how it works:

a. Policy Selection: Individuals interested in cryonics can choose a life insurance policy that aligns with their financial goals and cryonics arrangements. It is essential to select a policy that offers sufficient coverage to meet the anticipated cryonics expenses.

b. Beneficiary Designation: When purchasing a life insurance policy, individuals have the option to designate a cryonics organization or a trust as the beneficiary. By doing so, the insurance payout is directed towards covering cryonics expenses rather than being disbursed to traditional beneficiaries.

c. Premium Payments: Life insurance policies require regular premium payments. Individuals must ensure that they maintain the policy and make timely premium payments to keep the coverage active. Failure to pay premiums may result in the policy lapsing and the loss of coverage.

d. Collaboration with Cryonics Organization: It is crucial to collaborate with the chosen cryonics organization to ensure that the policy is structured appropriately. Cryonics organizations often have specific requirements and guidelines regarding beneficiary designation and policy assignment.

e. Policy Review and Updates: It is important to periodically review the life insurance policy to ensure that it aligns with the evolving cryonics arrangements. As cryonics expenses may change over time, adjusting the policy coverage and beneficiary designation may be necessary.

Cryonics Trusts:

Cryonics trusts are financial vehicles specifically designed to fund cryonics arrangements. These trusts are established to ensure that funds are allocated and managed appropriately for cryopreservation and related expenses. Here's how cryonics trusts work:

a. Trust Establishment: Cryonics trusts are typically created with the assistance of legal professionals specializing in trust formation. The trust document outlines the provisions for managing and distributing the trust funds for cryonics purposes.

b. Funding the Trust: Individuals can contribute funds to the cryonics trust over time, either through lump-sum contributions or periodic installments. These contributions are held and managed by a designated trustee or trustees.

c. Trustee Selection: The selection of a trustee is a critical aspect of cryonics trusts. Trustees are responsible for managing the trust funds and ensuring that they are used solely for cryonics expenses as specified in the trust document. Trustee selection should involve careful consideration of their experience, integrity, and ability to fulfill their fiduciary duties.

d. Trust Administration: Cryonics trusts require diligent administration and financial oversight. Trustees must keep accurate records, manage investments (if applicable), and ensure compliance with legal and regulatory requirements.

e. Collaboration with Cryonics Organization: Similar to life insurance, coordination with the chosen cryonics organization is crucial when establishing a cryonics trust. The organization can provide guidance on the appropriate structure and administration of the trust.

f. Periodic Review and Updates: Cryonics trusts should be reviewed periodically to ensure they align with current cryonics arrangements and legal requirements. Trust provisions may need to be updated to reflect changes in cryonics expenses, laws, or personal circumstances.

# Alternative Funding Methods

Cryonics, the practice of preserving the human body or brain at ultra-low temperatures with the hope of future revival and medical advancements, presents a unique set of challenges, including the cost of cryopreservation and associated expenses. While traditional funding methods such as life insurance and cryonics trusts are commonly used, there are alternative financial options available for individuals interested in pursuing cryonics. Below we will explore some alternative funding methods that can help individuals finance their cryonics arrangements.

Crowdfunding:

Crowdfunding has gained popularity in recent years as a means of raising funds for various purposes. Individuals interested in cryonics can leverage crowdfunding platforms to seek financial support from the public. Here's how it works:

a. Campaign Creation: Cryonics enthusiasts can create a compelling crowdfunding campaign explaining their motivations, goals, and the financial requirements of their cryonics arrangements. The campaign should highlight the potential benefits of cryonics and the impact it can have on the individual's life.

3. b. Outreach and Promotion: It is crucial to promote the crowdfunding campaign through various channels, such as social media, online communities, and cryonics-specific forums. Engaging with potential supporters and sharing the campaign widely can increase the chances of reaching the funding goal.

c. Offering Incentives: To encourage contributions, campaigners can offer incentives to donors, such as exclusive updates on their cryonics journey, personalized acknowledgments, or even the opportunity to attend cryonics-related events. These incentives can attract more supporters and increase the overall funding amount.

d. Transparency and Communication: Maintaining transparency throughout the crowdfunding campaign is essential. Regular updates on the progress of the cryonics arrangements and how the funds will be utilized can help build trust and keep supporters engaged.

Grant Programs and Scholarships:

In some cases, grant programs and scholarships may be available to support individuals pursuing cryonics. These funding opportunities can be offered by cryonics organizations, academic institutions, or private foundations with an interest in life extension research. Here's how individuals can explore these options:

a. Research and Applications: Individuals interested in cryonics should actively seek out grant programs and scholarships that support their cause. Conducting thorough research and identifying organizations or institutions with a focus on life extension and cryonics research can increase the chances of finding suitable funding opportunities.

b. Application Process: Grant programs and scholarships typically have specific application requirements and deadlines. Individuals should carefully review the guidelines and prepare a compelling application that highlights their passion for cryonics and their potential contributions to the field.

c. Networking and Collaborations: Building connections within the cryonics and life extension community can provide valuable insights into available grant programs and scholarship opportunities. Attending conferences, joining online forums, and engaging with researchers and experts in the field can open doors to funding possibilities.

Donations and Philanthropy:

Individuals interested in cryonics can seek financial support through donations from individuals or organizations passionate about life extension and advancing scientific frontiers. Here's how individuals can explore this option:

a. Identifying Potential Donors: Researching individuals or organizations that have shown an interest in cryonics, life extension, or related fields can provide leads for potential donors. These individuals or organizations may have a history of supporting scientific research or funding projects aligned with cryonics.

b. Building Relationships: Establishing personal connections and building relationships with potential donors is crucial. Individuals should engage in open and honest conversations, sharing their passion for cryonics and the potential impact it can have on future generations.

c. Presenting a Case for Support: When approaching potential donors, individuals should articulate their vision for cryonics, explaining how their cryonics arrangements align with the donor's

# The Cryonics Community: Support Networks and Advocacy

Imagine a world where death is not an inevitable end, but merely a temporary interruption. This is the vision of the cryonics community, a group of individuals who believe in the possibility of preserving human bodies or brains at low temperatures in the hope of future revival. Cryonics is a controversial and thought-provoking field that has gained attention over the years. However, what often goes unnoticed is the existence of vibrant support networks and advocacy efforts within the cryonics community. These networks provide emotional support, share scientific knowledge, and work tirelessly to promote the acceptance and advancement of cryonics.

At the heart of the cryonics community are the members themselves, individuals who have made the conscious decision to pursue cryopreservation. For many, this decision is driven by a desire to extend their lives, overcome incurable diseases, or simply to have a chance at a future where technology can reverse the effects of aging. The community serves as a haven for like-minded individuals, where they can connect with others who share their dreams, fears, and hopes. Through online forums, social media groups, and local meetups, cryonicists build strong bonds and find solace in knowing that they are not alone in their quest for a second chance at life.

Support networks within the cryonics community play a crucial role in providing emotional support and guidance. The decision to pursue cryonics is often met with skepticism and even ridicule from mainstream society. Family and friends may struggle to understand or accept this choice, which can create feelings of isolation and alienation. In response, cryonicists have developed support systems to bridge this gap. Online communities such as Cryonet, a long-standing cryonics mailing list, allow members to share their thoughts, concerns, and experiences. These platforms offer a safe space for cryonicists to express themselves, seek advice, and find encouragement from others who have walked a similar path.

Scientific knowledge and research are also central to the cryonics community. Cryonicists are driven by the belief that advances in science and technology will one day enable the revival of cryopreserved individuals. They stay up-to-date with the latest research in cryobiology, neurobiology, nanotechnology, and other relevant fields. Cryonicists actively engage in discussions, attend conferences, and contribute to the growing body of scientific literature. In doing so, they not only expand their own knowledge but also work towards legitimizing cryonics as a scientifically valid pursuit.

Advocacy is another crucial aspect of the cryonics community. Cryonicists are passionate about spreading awareness and countering misconceptions surrounding cryonics. They advocate for cryonics through various means, such as writing articles, giving talks, and engaging with the media. Organizations like the Cryonics Institute and Alcor Life Extension Foundation actively promote cryonics and provide educational resources to the public. These advocacy efforts aim to challenge societal biases and encourage open-minded discussions about the possibilities and ethical implications of cryopreservation.

In recent years, cryonics has garnered increased attention and support from notable individuals in the scientific and technological communities. Renowned scientists, such as Michio Kaku and Aubrey de Grey, have expressed cautious optimism about the potential of cryonics. High-profile figures, including Elon Musk and Peter Thiel, have publicly discussed their interest in cryopreservation as a means to extend human lifespan. This growing recognition has provided a boost to the cryonics community and its advocacy efforts, lending credibility to their cause.

The cryonics community also actively seeks collaborations with academia and the medical establishment. Researchers involved in cryobiology, neuroscience, and related fields are beginning to explore the potential applications of cryonics.

## Cryonics Conferences and Events

In the realm of cryonics, where the pursuit of extended life through the preservation of human bodies or brains is at the forefront, conferences and events play a vital role in fostering scientific exchange, community building, and advancing the field. These gatherings bring together cryonics enthusiasts, researchers, and experts from various disciplines to discuss cutting-edge technologies, share knowledge, and explore the ethical implications of cryopreservation. Below we will delve into the world of cryonics conferences and events, highlighting their significance in shaping the future of this fascinating field.

Cryonics conferences serve as a hub for scientific and technological advancements in the cryonics community. These events provide a platform for researchers and experts to present their findings, exchange ideas, and engage in thought-provoking discussions. Keynote speakers, renowned scientists, and pioneers in the field take the stage, delivering presentations on cryonics-related topics ranging from cryopreservation techniques to neurobiology and nanotechnology. By attending these conferences, participants gain insights into the latest research and breakthroughs, keeping them at the forefront of the rapidly evolving field.

One prominent cryonics conference is the annual "Alcor Conference on Life Extension," organized by the Alcor Life Extension Foundation. This event attracts cryonics advocates, professionals, and interested individuals from around the world. The conference features a diverse range of speakers, including cryonics researchers, biomedical engineers, and futurists, who share their expertise and vision for the future of cryonics. Workshops and panel discussions allow attendees to delve deeper into specific topics, fostering collaboration and the exchange of ideas.

The "Cryonics and Brain Preservation Conference" is another notable event that focuses on the scientific aspects of cryonics and brain preservation. This conference brings together scientists, neurologists, and experts in the field of cryobiology to discuss the latest developments in brain preservation techniques, neural imaging, and molecular repair. The event showcases cutting-edge research and encourages interdisciplinary collaborations between cryonics organizations and academic institutions.

Beyond scientific discussions, cryonics conferences also address the ethical, legal, and societal implications of cryopreservation. These aspects are explored in sessions dedicated to bioethics and public policy. Scholars and experts in these fields present their perspectives on the ethical considerations of cryonics, such as personal autonomy, consent, and the allocation of resources. The conferences provide a space for nuanced discussions and debates, helping to shape the ethical framework within which cryonics operates.

In addition to conferences, cryonics-related events take various forms, including symposiums, workshops, and seminars. These smaller-scale gatherings provide a more intimate setting for focused discussions and hands-on learning opportunities. For example, the "Cryonics Immersion Weekend" organized by the Cryonics Institute offers participants a chance to experience the cryopreservation process firsthand. Attendees can witness demonstrations, interact with cryonics professionals, and gain a deeper understanding of the intricate procedures involved in cryopreservation.

Cryonics events are not limited to academic and scientific circles. They also cater to the wider public, aiming to increase awareness and understanding of cryonics. Public lectures, TED-style talks, and media appearances by cryonics advocates serve as educational platforms to demystify cryonics and address common misconceptions. These events seek to engage the public in meaningful discussions about life extension, the potential impact of cryonics on society, and the ethical boundaries of pursuing immortality.

The internet has played a significant role in expanding the reach of cryonics conferences and events. Livestreaming and online platforms enable virtual participation, allowing individuals from all corners of the globe to access the wealth of knowledge shared at these gatherings.

# Online Communities and Social Media

In today's digital age, online communities and social media platforms have revolutionized the way we connect, share information, and engage in discussions. This holds true even in niche fields such as cryonics, where individuals with a shared interest in the preservation of human bodies or brains can come together and form vibrant online communities. These communities play a crucial role in fostering support, knowledge sharing, and advocacy for cryonics. Below we will explore the impact of online communities and social media platforms on the cryonics world.

One of the key benefits of online communities is their ability to connect individuals from different geographical locations who may not have access to local cryonics groups or organizations. Cryonics enthusiasts can now find solace and support through online forums, social media groups, and specialized websites. These platforms provide a safe space for individuals to share their experiences, concerns, and aspirations related to cryonics. They can engage in discussions, seek advice, and form meaningful connections with like-minded individuals who understand their perspectives and share their dreams of extended life.

The Cryonet mailing list is a prime example of an online community that has played a pivotal role in the cryonics world for decades. Cryonet serves as a platform for discussion among cryonicists, researchers, and interested individuals. It allows participants to exchange scientific insights, explore philosophical questions, and address practical concerns related to cryonics. Through Cryonet, the cryonics community has been able to form a global network of knowledge and support, transcending geographical boundaries.

Social media platforms have also become instrumental in connecting cryonicists and promoting awareness of cryonics to the broader public. Cryonics organizations and advocates utilize platforms like Facebook, Twitter, and Instagram to share news, scientific breakthroughs, and educational resources. These platforms serve as digital billboards for cryonics, attracting both curious individuals and potential supporters. Cryonics-related posts often spark discussions, debates, and the opportunity for advocates to dispel myths and misconceptions surrounding cryonics.

The reach and impact of social media are particularly evident during significant events or breakthroughs in cryonics. When a high-profile individual publicly expresses interest in cryonics or when new scientific findings related to cryopreservation emerge, social media platforms become abuzz with discussions, sharing of articles, and engagement from both supporters and skeptics. The rapid spread of information through social media not only increases public awareness but also influences the public perception and discourse surrounding cryonics.

Online communities and social media platforms also serve as a gateway to scientific knowledge and research in cryonics. Cryonicists actively share scientific papers, studies, and articles within these communities, creating a pool of collective intelligence. These platforms allow for the dissemination of research findings, fostering scientific discussions, and enabling collaboration between cryonics enthusiasts and professionals in related fields. As a result, advancements and discoveries can reach a wider audience, accelerating the progress of cryonics research.

Furthermore, online communities and social media have become indispensable tools for cryonics advocacy. Cryonicists and organizations leverage these platforms to share stories of individuals who have chosen cryopreservation, debunk common misconceptions, and engage with the public. Cryonics advocacy groups have established a strong presence online, utilizing informative websites, blogs, and social media accounts to educate the public about cryonics, address ethical concerns, and answer frequently asked questions.

The power of online communities and social media extends beyond the cryonics community itself. These platforms allow for collaborations between cryonics organizations and other entities, such as academic institutions and research facilities.

## Prominent Cryonicists and Their Stories

The world of cryonics is populated by a diverse array of individuals who have dedicated their lives to the pursuit of extended life through cryopreservation. Among them are prominent cryonicists whose stories inspire, educate, and shape the field. These individuals, through their passion, scientific contributions, and advocacy efforts, have played a significant role in advancing the understanding and acceptance of cryonics. Below we will explore the lives and achievements of some of these prominent cryonicists.

Robert Ettinger:

Often referred to as the "Father of Cryonics," Robert Ettinger is considered one of the pioneers in the field. In 1962, he published "The Prospect of Immortality," a groundbreaking book that laid the foundation for cryonics as a scientific and philosophical pursuit. Ettinger's work popularized the concept of cryonics and sparked discussions on life extension and the possibilities of future revival.

Fred Chamberlain:

Fred Chamberlain, a key figure in cryonics history, played a crucial role in establishing the first cryonics organization, the Cryonics Society of California (CSC). Chamberlain co-founded CSC in 1966, making it the first organization dedicated to cryonics. His contributions helped pave the way for the growth of cryonics as a legitimate field of scientific inquiry.

Saul Kent:

Saul Kent, a prominent advocate for cryonics, has been instrumental in raising public awareness about the practice. As the co-founder of the Life Extension Foundation, Kent has worked tirelessly to promote cryonics as a viable option for life extension. His efforts have included supporting research initiatives, organizing conferences, and providing educational resources to the cryonics community.

Michael Darwin:

Michael Darwin, a key figure in cryonics history, has made significant contributions to cryonics research and organization. Darwin has worked as a researcher, technician, and consultant in various cryonics organizations. His technical expertise and insights have helped shape the cryopreservation process, improving the chances of successful preservation and future revival.

Max More:

Max More, a prominent transhumanist philosopher and cryonics advocate, has been actively involved in the cryonics community for many years. More has contributed to the philosophical underpinnings of cryonics, emphasizing the importance of individual choice, personal identity, and the potential benefits of future technologies. He has also held leadership positions in cryonics organizations and has been an influential voice in shaping the public perception of cryonics.

Ralph Merkle:

Ralph Merkle, a renowned computer scientist and cryonics advocate, has made significant contributions to the field. He is credited with developing the concept of cryonics nanotechnology, which explores the use of molecular-level technologies for repairing and rejuvenating cryopreserved tissue. Merkle's work has pushed the boundaries of cryonics research and provided a vision for future advancements in the field.

Linda Chamberlain:

Linda Chamberlain, the widow of Fred Chamberlain, has been a prominent figure in cryonics advocacy. Following her husband's passing, Chamberlain took on leadership roles within cryonics organizations, working tirelessly to promote awareness, support research initiatives, and provide guidance to the cryonics community. Her dedication has helped sustain and advance the field of cryonics.

Aschwin de Wolf:

Aschwin de Wolf, a cryonics researcher and co-founder of Advanced Neural Biosciences, has made significant contributions to cryonics through his scientific expertise. De Wolf has focused on improving cryopreservation techniques, particularly in the field of brain preservation.

# Preparing for Cryopreservation: Legal, Medical, and Personal Considerations

Cryopreservation, the process of preserving human bodies or brains at ultra-low temperatures with the hope of future revival, raises a host of legal, medical, and personal considerations. As individuals explore the possibility of cryonics as an option for extending their lives, it is crucial to navigate these complex aspects. Below we will delve into the important considerations one must take into account when preparing for cryopreservation.

From a legal perspective, several key considerations come into play. First and foremost, it is crucial to ensure that cryopreservation arrangements are legally recognized and enforceable. Engaging legal professionals with expertise in cryonics is advisable to ensure that appropriate documentation, such as a cryopreservation contract, is in place. These legal agreements outline the individual's wishes for cryopreservation, specify the responsibilities of cryonics organizations or service providers, and address any financial or property-related matters. It is essential to consult local laws and regulations to ensure compliance and to guarantee that the cryopreservation process will proceed smoothly.

Medical considerations are also critical when preparing for cryopreservation. It is essential to establish a relationship with healthcare professionals who are knowledgeable about and supportive of the cryonics process. This may involve finding a cryonics-friendly doctor who is willing to assist in the necessary procedures before and after legal death. Establishing this relationship early on is crucial, as it ensures that medical personnel are informed about the individual's cryopreservation wishes and can provide the necessary cooperation and assistance when the time comes. Collaborating with medical professionals who understand the protocols involved in cryopreservation can help mitigate potential complications and ensure a smooth transition into the cryopreservation process.

Personal considerations are equally significant when preparing for cryopreservation. The decision to pursue cryonics is deeply personal and often requires careful reflection. It is important to consider the emotional and psychological aspects of embarking on this journey. Open and honest communication with loved ones is essential, as they may play a crucial role in supporting the individual's wishes and ensuring that the necessary arrangements are made. Engaging in discussions with family members and friends about cryonics can help foster understanding, address concerns, and ensure that the individual's desires are respected.

Financial planning is another critical aspect of preparing for cryopreservation. Cryonics is a long-term commitment that requires financial resources. Individuals considering cryopreservation should carefully assess the costs involved, including the initial cryopreservation fees and ongoing membership fees for cryonics organizations. Creating a financial plan that ensures the availability of funds for cryopreservation is essential. Some individuals choose to establish trust funds, life insurance policies, or other financial mechanisms specifically designated for cryonics expenses. Engaging with financial advisors who are knowledgeable about cryonics can provide valuable guidance and assistance in developing a sound financial plan.

Additionally, individuals preparing for cryopreservation should consider the logistics involved in the process. This includes arranging for the transportation of the body or brain to the cryonics facility, either domestically or internationally. Cooperation and coordination with the cryonics organization or service provider are crucial to ensure a seamless transition and proper handling of the remains. Advance planning and communication with the chosen cryonics organization can help address any concerns or questions regarding transportation logistics, ensuring that the process is carried out according to the individual's wishes.

# Ensuring Proper Documentation and Consent

Cryonics, the practice of preserving human bodies or brains at ultra-low temperatures with the hope of future revival, necessitates meticulous attention to documentation and consent. Given the unique nature of cryonics and the ethical considerations involved, ensuring that appropriate legal and consent-related measures are in place is of utmost importance. Below we will explore the significance of proper documentation and consent in the context of cryonics.

One crucial aspect of ensuring proper documentation in cryonics is the establishment of legal frameworks that recognize and protect the rights and wishes of individuals seeking cryopreservation. This involves creating legally binding documents, such as a cryonics contract or a cryopreservation agreement, which clearly outline the individual's desires and intentions regarding cryonics. These documents specify the individual's consent for cryopreservation, delineate the responsibilities of cryonics organizations or service providers, and address any financial or property-related matters. Engaging legal professionals with expertise in cryonics is advisable to ensure that the documentation is legally enforceable and compliant with local laws and regulations.

Obtaining informed consent is a fundamental requirement in cryonics. Informed consent means that individuals fully understand the nature of cryonics, including the processes involved, the potential risks, and the uncertain outcomes. Cryonics organizations have a responsibility to provide comprehensive information to potential cryonics patients, ensuring that they are well-informed about the procedure, its limitations, and the potential challenges associated with future revival. This may involve providing educational materials, organizing informational sessions, or offering opportunities for individuals to engage in discussions with cryonics professionals. Informed consent empowers individuals to make well-informed decisions about cryonics and ensures that their wishes are respected.

In addition to the legal and consent-related documents, it is crucial to establish a system for the secure storage and retrieval of these documents. Cryonics organizations must have robust record-keeping protocols in place to ensure the proper storage and accessibility of cryonics-related documentation. This includes securely archiving consent forms, contracts, and any other relevant legal or medical documents. Additionally, cryonics organizations should implement procedures that allow for the timely retrieval of documentation when needed. This ensures that the individual's wishes and legal rights are upheld and facilitates a smooth transition into the cryopreservation process.

Transparency and accountability are key principles in cryonics, and proper documentation plays a vital role in upholding these values. Cryonics organizations must be transparent about their practices, policies, and the potential challenges associated with cryopreservation. This transparency extends to documenting the procedures involved in cryopreservation, including the steps taken before, during, and after the preservation process. Detailed records allow for accountability and enable the evaluation and improvement of cryopreservation protocols over time.

Furthermore, ensuring the authenticity and validity of documentation is essential. Cryonics organizations must implement robust verification mechanisms to confirm the authenticity of consent and legal documents. This may involve the use of notaries, witnesses, or other legally recognized methods of authentication. Verifying the authenticity of documentation adds an extra layer of security and ensures that the individual's wishes are accurately represented and upheld.

Beyond the legal and technical considerations, communication and ongoing dialogue with loved ones are crucial when it comes to documentation and consent in cryonics. Individuals who choose cryopreservation should engage in open and honest conversations with their family members, friends, and healthcare providers. These discussions help ensure that their wishes regarding cryopreservation are well-known and understood by those closest to them. Clear communication fosters trust, minimizes the potential for conflicts or misunderstandings, and provides a support network for the individual seeking cryopreservation.

# Navigating Medical and End-of-Life Care

Cryonics, the practice of preserving human bodies or brains at ultra-low temperatures with the hope of future revival, raises unique considerations regarding medical and end-of-life care. Individuals who choose cryopreservation must navigate medical decisions, collaborate with healthcare professionals, and ensure that their cryonics wishes are respected during their end-of-life journey. Below we will explore the challenges and considerations associated with medical and end-of-life care in the context of cryonics.

When considering cryonics, individuals should engage in open and honest discussions with their healthcare providers. It is crucial to find medical professionals who are knowledgeable about and supportive of cryonics. Establishing a cooperative relationship with a cryonics-friendly doctor is essential to ensure that the necessary medical procedures before and after legal death are carried out in alignment with cryopreservation requirements. This includes coordinating with medical professionals to ensure that procedures such as cardiopulmonary support, medications, and rapid cooling are appropriately implemented to maximize the potential for successful cryopreservation.

Communication and coordination with healthcare providers should not be limited to the individual's personal physician. It is important to communicate cryonics wishes with emergency medical services (EMS) personnel, hospital staff, and other healthcare professionals who may be involved in end-of-life care. In case of an emergency, having a clear, concise, and easily accessible document, such as a medical directive or a "Do Not Resuscitate" (DNR) order, that outlines the individual's cryonics wishes can help ensure that medical interventions are aligned with their end-of-life plans.

Planning for end-of-life care requires considering the circumstances under which cryopreservation will occur. Since cryonics typically takes place after legal death is declared, it is important to have a clear understanding of the legal requirements and procedures involved in pronouncing death. Collaborating with cryonics organizations and healthcare professionals who are familiar with the legal and medical aspects of cryopreservation can help navigate the complexities associated with end-of-life care.

It is crucial to involve family members or trusted individuals in the process of end-of-life decision-making. Open communication with loved ones about cryonics, its rationale, and the individual's wishes is essential. Including family members in discussions with healthcare providers can help ensure that their concerns are addressed, and that they understand and support the individual's choice for cryopreservation. This can alleviate potential conflicts and facilitate a smoother transition during the end-of-life journey.

Advance care planning is another crucial aspect of navigating medical and end-of-life care in the context of cryonics. This involves documenting the individual's preferences for medical interventions, such as the use of life-sustaining measures, pain management, and organ donation. Creating an advance directive or living will that specifically addresses cryonics-related wishes can provide guidance to healthcare providers and ensure that the individual's desires are respected. Reviewing and updating these documents regularly is important to reflect changes in preferences or legal requirements.

In addition to medical considerations, financial planning is crucial when it comes to end-of-life care and cryonics. Cryopreservation is a long-term commitment that requires financial resources. Individuals pursuing cryonics should consider the costs associated with cryopreservation, ongoing membership fees for cryonics organizations, and any additional expenses related to medical procedures and legal documentation. Engaging with financial advisors who are knowledgeable about cryonics can help develop a sound financial plan to ensure the availability of funds for end-of-life care and cryopreservation expenses.

# Communicating Your Wishes to Family and Friends

When considering cryonics, the practice of preserving human bodies or brains at ultra-low temperatures with the hope of future revival, one of the crucial aspects to navigate is effectively communicating your wishes to family and friends. Sharing your decision to pursue cryopreservation with loved ones requires careful thought, open dialogue, and empathy. Below we will explore the importance of communicating your cryonics wishes, the challenges that may arise, and strategies for fostering understanding and support among family and friends.

First and foremost, it is important to approach the conversation with empathy and respect for differing perspectives. Cryonics is a concept that may be unfamiliar and even unsettling to some people. Understand that your loved ones may have concerns, reservations, or questions about cryonics. By acknowledging their emotions and addressing their concerns, you create an environment where open dialogue can take place.

Begin by educating yourself about cryonics and its scientific basis. Gathering accurate information about the process, its history, and the potential outcomes will enable you to articulate your views more effectively. This knowledge will help you address common misconceptions and provide evidence-based explanations for your decision to pursue cryonics. Presenting the information in a calm and rational manner can help alleviate fears and misconceptions.

Consider sharing resources such as books, articles, documentaries, or online materials that provide objective and accurate information about cryonics. These resources can help your loved ones gain a deeper understanding of the scientific principles and ethical considerations behind cryonics. Encourage them to explore the materials at their own pace and offer to answer any questions or concerns that may arise.

Timing is crucial when broaching the topic of cryonics with family and friends. Choose a moment when everyone is relaxed and receptive, ensuring that there is sufficient time for a meaningful conversation. It may be helpful to set aside dedicated time for the discussion, allowing for uninterrupted dialogue.

Express your motivations and reasons for considering cryonics. Share your aspirations for extended life, the potential benefits you see in the future, and your desire to take advantage of advancements in medical technology. Personalizing your perspective can help your loved ones understand the emotional and philosophical underpinnings of your decision.

Active listening is a key component of effective communication. Encourage your loved ones to express their thoughts, concerns, and fears openly. Validate their emotions and demonstrate that you value their perspectives. By actively listening, you create space for mutual understanding and respect.

Recognize that not everyone may immediately embrace or support your decision. It is important to remain patient and understanding. Give your loved ones time to process the information and reflect on their own beliefs. Remember that their initial reactions may stem from a place of concern or fear for your well-being.

In some cases, involving a neutral third party can be beneficial. This may involve arranging a meeting with a healthcare professional, a counselor, or a cryonics advocate who can provide an objective perspective and address any medical, ethical, or psychological questions that arise. These discussions can help your loved ones gain insight and reassurance from experts in the field.

Consider developing a comprehensive plan that outlines your wishes for end-of-life care and cryonics. This may include creating legal documents such as a living will, medical directive, or power of attorney that explicitly address your cryonics desires. Share these documents with your loved ones, ensuring that they understand their role and responsibilities in honoring your wishes.

Additionally, it can be helpful to involve a trusted family member or friend in the decision-making process. Designating a "healthcare proxy" or "agent" who understands your desires and is committed to advocating for your cryonics wishes can provide a sense of reassurance to both you and your loved ones.

# The Future of Cryonics: Possibilities and Challenges

Cryonics, the practice of preserving human bodies or brains at ultra-low temperatures with the hope of future revival, has long been a subject of fascination and debate. As the field continues to evolve, the future of cryonics holds both exciting possibilities and significant challenges. Below we will explore the potential advancements, ethical considerations, and scientific hurdles that shape the future of cryonics.

Advancements in scientific and medical technologies offer promising possibilities for the future of cryonics. Researchers are continuously exploring novel approaches to cryopreservation and tissue preservation. Cryobiologists are developing improved techniques that minimize cellular damage during the freezing process, such as the use of cryoprotectants and vitrification. These advancements aim to enhance the preservation of biological structures and increase the likelihood of successful revival in the future.

The field of regenerative medicine also holds great potential for cryonics. Stem cell research, tissue engineering, and organ transplantation advancements may play a crucial role in repairing and rejuvenating cryopreserved tissues. By harnessing the power of regenerative medicine, scientists envision a future where damaged organs and tissues can be restored, paving the way for successful revival of cryopreserved individuals.

Nanotechnology is another field that could revolutionize cryonics. Researchers are exploring the use of nanomaterials and nanorobots for cellular repair and molecular-level interventions. Nanotechnology could potentially enable precise manipulation of cryopreserved tissues, repair of cellular damage, and even targeted interventions at the molecular level. These advancements may significantly enhance the chances of successful revival and restoration of cryopreserved individuals.

In addition to scientific advancements, societal attitudes towards cryonics may change in the future. As cryonics becomes more widely known and accepted, public perception may shift, leading to increased support and investment in the field. High-profile figures, including scientists, entrepreneurs, and celebrities, have already expressed interest in cryonics, bringing attention and credibility to the practice. With continued advocacy and education, cryonics may gain broader acceptance as a legitimate option for extending human life.

However, the future of cryonics also presents significant challenges that need to be addressed. One of the primary challenges is the cost associated with cryopreservation and long-term storage. Cryonics is currently an expensive endeavor, requiring substantial financial resources to cover initial cryopreservation fees, ongoing membership fees, and the cost of maintaining cryonics organizations and facilities. Making cryonics more accessible and affordable will be essential for its widespread adoption and long-term viability.

Ethical considerations surrounding cryonics also require careful examination. Questions regarding personal identity, the allocation of resources, and consent in future revival scenarios are complex and multifaceted. Society will need to engage in thoughtful discussions and ethical debates to address these issues and establish guidelines and regulations that protect the rights and interests of cryopreserved individuals.

Another challenge lies in the unknowns surrounding the future state of technology and society. Cryopreservation is a long-term commitment, and the revival of cryopreserved individuals relies on advancements in science and medicine that may not yet exist. Uncertainties regarding the feasibility and timeline of future revival pose both practical and existential challenges for cryonics. However, proponents of cryonics argue that the potential benefits of extended life outweigh the uncertainties, emphasizing the importance of embracing the unknown for the chance at future revival.

Public education and outreach will continue to be crucial in shaping the future of cryonics. Informing the general public about the scientific principles, advancements, and potential benefits of cryonics will be vital in dispelling misconceptions and fostering informed discussions.

# Technological Breakthroughs on the Horizon

Cryonics, the practice of preserving human bodies or brains at ultra-low temperatures with the hope of future revival, is an ever-evolving field that constantly pushes the boundaries of science and technology. As researchers and cryonicists strive for advancements in the field, several technological breakthroughs are on the horizon. These breakthroughs have the potential to revolutionize cryonics, improving preservation techniques and increasing the chances of successful revival. Below we will explore some of the technological advancements that hold promise for the future of cryonics.

One of the areas that researchers are actively exploring is the development of improved cryoprotectants and vitrification techniques. Cryoprotectants are substances that help protect cells and tissues from damage during the freezing process. Scientists are working to refine existing cryoprotectants and discover new ones that can better preserve the structural integrity of cells and minimize ice formation. The goal is to find cryoprotectants that are more efficient, less toxic, and capable of achieving vitrification—a state in which tissues solidify without forming ice crystals. Advancements in cryoprotectants and vitrification techniques would significantly enhance the preservation process and increase the chances of successful revival.

Nanotechnology is another field that holds tremendous potential for cryonics. Researchers envision using nanoscale materials and devices to repair and rejuvenate cryopreserved tissues. Nanorobots, for example, could be designed to navigate through the body, identify damaged cells or structures, and carry out repairs at the molecular level. These nanorobots could repair cellular damage, remove toxic substances, or introduce regenerative factors to restore functionality. While nanotechnology is still in its early stages, ongoing research and breakthroughs in the field may pave the way for advanced interventions in cryonics.

Advancements in regenerative medicine and tissue engineering are also poised to impact the future of cryonics. Scientists are exploring methods to regenerate damaged or degenerated tissues, organs, and even entire bodies. Stem cell research, for instance, offers the potential to generate new cells and tissues that can replace damaged ones. Techniques such as induced pluripotent stem cells (iPSCs) allow for the reprogramming of adult cells to a pluripotent state, enabling them to differentiate into different cell types. This breakthrough in stem cell technology could potentially be utilized to repair and replace damaged tissues in cryopreserved individuals upon future revival.

Artificial intelligence (AI) and machine learning also have the potential to revolutionize cryonics. These technologies can assist in the analysis and interpretation of complex biological data, aiding researchers in understanding the intricacies of cryopreservation and revival processes. AI algorithms can help identify patterns, optimize cryopreservation protocols, and assist in the analysis of large datasets. Machine learning techniques can also facilitate personalized medicine approaches, enabling tailored interventions and treatments for cryopreserved individuals upon future revival.

Advancements in imaging technologies are poised to enhance the preservation and revival process in cryonics. High-resolution imaging techniques, such as electron microscopy and nanoscale imaging, can provide detailed information about the structure and composition of cryopreserved tissues. These imaging technologies allow for a comprehensive understanding of cellular organization and enable researchers to assess the effectiveness of preservation methods and evaluate tissue integrity. Furthermore, advancements in neural imaging techniques can potentially provide insights into the connectivity and functionality of cryopreserved brains, aiding in the restoration of neural networks upon revival.

The field of cryobiology is constantly evolving, with researchers exploring innovative approaches to improve cryopreservation techniques. Cryopreservation using helium-based cooling, known as helium vitrification, is an emerging technology that shows promise.

# Overcoming Societal and Regulatory Hurdles

Cryonics, the practice of preserving human bodies or brains at ultra-low temperatures with the hope of future revival, faces a range of societal and regulatory hurdles. As a field that challenges traditional notions of death, longevity, and medical practices, cryonics often encounters skepticism, legal constraints, and ethical concerns. Below we will explore the societal and regulatory hurdles that cryonics must overcome and the potential strategies to address them.

One of the primary societal hurdles for cryonics is the skepticism and misunderstanding surrounding the practice. Cryonics challenges deeply ingrained beliefs about death and the limits of medical science. Many people find it difficult to envision a future where cryopreserved individuals can be revived and restored to full health. Overcoming this hurdle requires increased public education and outreach efforts. Cryonics organizations can engage in community events, public lectures, and media appearances to present accurate information about the science, research, and potential benefits of cryonics. Sharing success stories, highlighting scientific advancements, and dispelling common misconceptions can help bridge the knowledge gap and foster more informed public discussions.

Cryonics also faces regulatory hurdles due to its unique nature. The legal and regulatory frameworks in many jurisdictions are often not designed to accommodate the complexities of cryonics. For instance, issues related to legal death, property rights, and inheritance laws may arise when determining the status and treatment of cryopreserved individuals. To overcome these hurdles, cryonics organizations and advocates must engage with policymakers, legal experts, and bioethicists to address the legal and regulatory gaps. Collaborative efforts can lead to the development of specific legislation and regulations that protect the rights and interests of cryopreserved individuals, establish clear guidelines for end-of-life care, and recognize the validity of cryonics contracts and consent.

Another societal challenge is the affordability and accessibility of cryonics. Cryopreservation is currently an expensive undertaking, with significant upfront costs and ongoing membership fees. This financial barrier limits access to cryonics for many individuals. To overcome this hurdle, cryonics organizations can explore innovative financing options, such as insurance plans or crowdfunding initiatives, to make cryonics more affordable and accessible. Collaborative efforts between cryonics organizations and medical insurance providers may also help establish insurance coverage for cryonics procedures, reducing the financial burden on individuals and families.

Ethical concerns surrounding cryonics present yet another hurdle. Questions related to personal identity, consent, and the allocation of resources often arise in discussions about cryonics. Critics argue that cryonics diverts resources from more immediate healthcare needs and that the future revival of cryopreserved individuals may raise ethical dilemmas. To address these concerns, cryonics organizations and advocates can engage in ethical discussions, collaborate with bioethicists, and develop guidelines that uphold ethical principles. Emphasizing the voluntary nature of cryonics, the potential benefits to individuals and society, and the commitment to ongoing scientific research and responsible practices can help alleviate ethical concerns and foster a more nuanced understanding of the field.

Cultural and religious beliefs also pose challenges for cryonics acceptance. Different cultures and religions have diverse perspectives on death, afterlife, and the sanctity of the human body. Some individuals may find cryonics incompatible with their religious or cultural beliefs. Overcoming this hurdle requires sensitivity and respect for diverse belief systems. Engaging in interfaith dialogues, collaborating with religious leaders, and facilitating discussions that explore the compatibility of cryonics with various belief systems can help address these concerns and promote a more inclusive understanding of cryonics.

# Cryonics and the Quest for Immortality

Cryonics, the practice of preserving human bodies or brains at ultra-low temperatures with the hope of future revival, is often intertwined with the human quest for immortality. The idea of achieving indefinite life extension through cryopreservation has captivated the imagination of many individuals. Below we will explore the connection between cryonics and the quest for immortality, delving into the motivations, challenges, and philosophical implications associated with this pursuit.

At the heart of the quest for immortality is the desire to transcend the limitations of human mortality. Throughout history, humans have sought ways to overcome aging, disease, and death. Cryonics offers a unique approach to this age-old quest by preserving the body or brain at the point of legal death, with the hope that future advancements in science and technology will enable revival and restoration to life.

One of the primary motivations for pursuing cryonics is the belief that death is not an inevitable and irreversible outcome. Proponents of cryonics argue that death should be viewed as a medical condition that can potentially be treated and overcome. They believe that future breakthroughs in fields such as regenerative medicine, nanotechnology, and artificial intelligence could enable the repair and rejuvenation of cryopreserved individuals, effectively extending their lives indefinitely.

The quest for immortality through cryonics, however, is not without its challenges. Scientific and technological uncertainties pose significant hurdles. While cryonics organizations have made strides in improving preservation techniques, the process of vitrification—the solidification of tissues without ice formation—is not yet perfected. Cellular damage and ice crystal formation during the freezing process can present obstacles to successful revival. Overcoming these challenges requires ongoing scientific research and advancements in cryopreservation methods.

Financial considerations also play a crucial role in the quest for immortality through cryonics. Cryopreservation and long-term storage come with significant costs. Individuals interested in cryonics must carefully plan for these expenses, including initial cryopreservation fees, ongoing membership fees, and the financial resources needed to sustain cryonics organizations and facilities over extended periods of time. Access to adequate funding and affordable cryonics options can impact the feasibility and accessibility of pursuing cryonics as a means of achieving immortality.

The quest for immortality through cryonics raises profound philosophical questions about personal identity, the nature of consciousness, and the implications of extended life. Critics argue that cryonics may not preserve the continuity of personal identity, as the revival of cryopreserved individuals may result in altered memories, personality, or subjective experience. They question whether the revived individual would truly be the same person who underwent cryopreservation. These philosophical concerns prompt us to reflect on the fundamental nature of personal identity and the complex relationship between mind and body.

Ethical considerations also emerge in the quest for immortality through cryonics. Critics argue that cryonics diverts resources from more immediate healthcare needs and that prioritizing the revival of cryopreserved individuals raises ethical dilemmas. Questions regarding the allocation of resources, social equity, and the distribution of life-extending technologies come to the forefront. Engaging in ethical discussions and addressing these concerns is essential in navigating the ethical landscape surrounding cryonics and its implications for the pursuit of immortality.

While cryonics offers a potential pathway to extended life, it is important to recognize that achieving immortality in the literal sense may remain an elusive goal. The quest for immortality through cryonics is multifaceted and encompasses scientific, technological, philosophical, and ethical dimensions. It challenges us to confront fundamental questions about life, death, and the human condition.

# The Intersection of Cryonics and Artificial Intelligence

Cryonics, the practice of preserving human bodies or brains at ultra-low temperatures with the hope of future revival, and artificial intelligence (AI), the field of computer science focused on creating intelligent machines, may seem like disparate realms. However, these two fields intersect in intriguing ways, presenting opportunities and challenges that shape the future of cryonics. Below we will explore the relationship between cryonics and artificial intelligence, examining how AI can potentially enhance cryonics practices and the implications of this intersection.

Artificial intelligence plays a crucial role in the advancement of cryonics research and preservation techniques. AI algorithms can aid in the analysis and interpretation of complex biological data, assisting researchers in understanding the intricacies of cryopreservation and revival processes. By analyzing large datasets, AI can identify patterns, optimize cryopreservation protocols, and improve the quality and efficiency of preservation methods. Machine learning techniques can also contribute to personalized medicine approaches, facilitating tailored interventions and treatments for cryopreserved individuals upon future revival.

Furthermore, AI can assist in the revival and restoration process of cryopreserved individuals. Advanced AI algorithms can aid in the reconstruction of damaged neural networks and the restoration of brain functionality. These algorithms can analyze the structural and functional data obtained from the cryopreserved brain and simulate the neural connections, potentially enabling the revival of cognitive functions and memories. AI-based models can help bridge the gaps in knowledge and understanding of the brain, guiding scientists in the complex task of restoring consciousness and identity in cryonics.

Another area where cryonics and AI intersect is in the field of mind uploading. Mind uploading refers to the transfer of a person's consciousness, memories, and identity into a digital substrate, such as a computer simulation or an artificial body. This concept merges cryonics with AI, as the preservation of the brain through cryonics can potentially provide a starting point for the process of mind uploading. AI technologies, particularly in the realm of cognitive neuroscience and brain-computer interfaces, play a pivotal role in developing the tools and methodologies required for successful mind uploading.

While the intersection of cryonics and AI presents exciting possibilities, it also raises ethical and philosophical questions. One of the primary concerns is the preservation of personal identity. Critics argue that the revival of cryopreserved individuals, even with the aid of AI, may result in altered memories, personality, or subjective experience. They question whether the revived individual would maintain the continuity of personal identity and be the same person who underwent cryopreservation. Addressing these concerns requires a careful examination of the relationship between consciousness, identity, and the potential impact of AI-based revival on personal experiences.

Ethical considerations also emerge at the intersection of cryonics and AI. Questions regarding consent, autonomy, and the use of AI in decision-making processes arise. It is crucial to ensure that the integration of AI in cryonics practices respects individual autonomy and aligns with the individual's wishes. Ethical frameworks must be developed to guide the responsible use of AI in the context of cryonics, addressing concerns related to the potential manipulation or alteration of personal identity, privacy issues, and the equitable distribution of resources and opportunities associated with AI-based revival.

In addition, societal implications come into play. The widespread use of AI in cryonics practices can lead to social inequalities, with access to advanced AI technologies potentially becoming a determining factor in who can afford and benefit from cryonics procedures. Ensuring equitable access to both cryonics and AI technologies is crucial to avoid exacerbating existing disparities in healthcare and longevity opportunities. Education and public outreach efforts are necessary to foster public understanding and engagement with the societal implications of the intersection of cryonics and AI.

## AI in the Cryopreservation Process

Cryonics, the practice of preserving human bodies or brains at ultra-low temperatures with the hope of future revival, has long relied on advancements in science and technology. As the field continues to evolve, one area that shows great promise is the integration of artificial intelligence (AI) in the cryopreservation process. AI algorithms and techniques offer new possibilities to enhance and optimize cryonics practices. Below we will explore the role of AI in the cryopreservation process and the potential benefits it can bring.

One of the primary areas where AI can contribute to cryonics is in the optimization of cryopreservation protocols. Cryopreservation involves cooling the body or brain to extremely low temperatures to halt biological processes. However, achieving optimal cooling rates and minimizing damage to tissues during the freezing process is a complex task. AI algorithms can analyze vast amounts of data collected from previous cryopreservation procedures and identify patterns and correlations between cooling rates, cryoprotectant concentrations, and tissue damage. By learning from this data, AI can assist in the development of more effective and efficient cryopreservation protocols, ultimately improving the preservation quality and increasing the chances of successful revival.

Furthermore, AI can aid in the real-time monitoring and control of cryopreservation processes. Temperature gradients, cryoprotectant concentrations, and other variables need to be carefully monitored and adjusted during the freezing and storage stages of cryopreservation. AI algorithms can analyze sensor data, predict trends, and make informed decisions to optimize these parameters. By continuously monitoring and adjusting the cryopreservation environment, AI can ensure that optimal conditions are maintained throughout the process, minimizing the risk of damage and improving the overall preservation outcome.

Another area where AI can play a significant role is in the analysis and interpretation of cryopreserved tissue samples. Cryonics organizations often collect samples for future analysis, hoping to gain insights into the cellular structure, molecular composition, and potential damage caused during cryopreservation. Analyzing these samples manually can be a time-consuming and labor-intensive task. AI algorithms, however, can automate this process, rapidly analyzing large datasets and identifying cellular structures, patterns, and abnormalities. The use of AI in tissue analysis can accelerate research efforts, allowing for a better understanding of cryopreservation outcomes and guiding improvements in preservation techniques.

Moreover, AI has the potential to contribute to the future revival process of cryopreserved individuals. As technologies advance, scientists are exploring the possibility of using AI to assist in the reconstruction of damaged neural networks. AI algorithms can analyze the structural and functional data obtained from cryopreserved brains, simulate neural connections, and assist in the restoration of brain functionality. This integration of AI in the revival process holds promise for improving the chances of successfully restoring cognitive functions and memories in cryopreserved individuals.

The integration of AI in cryonics practices also raises ethical and philosophical questions. One of the primary concerns is the preservation of personal identity. Critics argue that the revival of cryopreserved individuals with the aid of AI may result in altered memories, personality, or subjective experience. They question whether the revived individual would maintain the continuity of personal identity. Addressing these concerns requires careful consideration and ethical frameworks that respect individual autonomy and ensure that the integration of AI aligns with the individual's wishes.

Furthermore, AI in cryonics highlights the importance of responsible and transparent implementation. The use of AI algorithms must be subject to rigorous testing, validation, and transparency to ensure the accuracy and reliability of the results. The integration of AI should also be accompanied by robust regulations and guidelines that address issues such as data privacy, informed consent, and the equitable distribution of AI-based cryonics technologies.

# AI-Assisted Cryonics Research and Development

Cryonics, the practice of preserving human bodies or brains at ultra-low temperatures with the hope of future revival, is a field that continually seeks advancements in research and development. Artificial intelligence (AI) has emerged as a powerful tool in various scientific disciplines, and its integration into cryonics research shows great promise. Below we will explore the role of AI in assisting cryonics research and development, highlighting its potential benefits and implications for the future of the field.

One area where AI can significantly contribute to cryonics research is in the analysis and interpretation of complex biological data. Cryonics organizations and researchers collect vast amounts of data during the cryopreservation process, including images, tissue samples, and biochemical measurements. Analyzing and extracting meaningful insights from this data manually can be time-consuming and labor-intensive. AI algorithms, however, can process and analyze large datasets with remarkable speed and accuracy. Machine learning techniques can detect patterns, identify correlations, and recognize anomalies that may otherwise go unnoticed. By applying AI to cryonics research, scientists can gain valuable insights into the effects of cryopreservation on cellular structures, identify factors that influence preservation quality, and develop strategies to improve cryopreservation techniques.

AI can also play a crucial role in optimizing cryopreservation protocols. The freezing and preservation of biological tissues involve intricate processes and numerous variables, including cooling rates, cryoprotectant concentrations, and storage conditions. Determining the optimal parameters to achieve the highest quality of preservation can be challenging. AI algorithms can analyze large datasets containing information on preservation outcomes, tissue properties, and experimental conditions. By leveraging this data, AI can identify patterns and correlations, enabling researchers to refine and optimize cryopreservation protocols. This optimization process can lead to improved preservation outcomes, enhancing the chances of successful revival in the future.

Moreover, AI can assist in the development of innovative technologies and interventions that enhance cryopreservation processes. The integration of AI with other emerging fields, such as nanotechnology and robotics, opens up new possibilities for cryonics research and development. AI algorithms can aid in the design and control of nanorobots or nanoscale devices used for cellular repair and molecular interventions. These AI-assisted technologies can precisely target damaged tissues, repair cellular structures, and remove toxic substances, ultimately improving the preservation and revival potential of cryonics procedures.

Another area where AI can make a significant impact is in predictive modeling and simulation. AI algorithms can analyze large datasets of cryopreserved tissue properties and outcomes, as well as other relevant biological data, to create predictive models. These models can simulate the effects of different preservation conditions, predict the likelihood of tissue damage, and even estimate the potential outcomes of future revival attempts. By using AI-driven simulations, researchers can make informed decisions, optimize revival strategies, and reduce experimental risks and costs.

The integration of AI in cryonics research and development also raises ethical and philosophical considerations. One primary concern is the impact of AI on personal identity. Critics argue that the use of AI in the revival of cryopreserved individuals may result in altered memories, personality, or subjective experiences. The question of whether the revived individual would maintain the continuity of personal identity remains a complex philosophical issue. Ethical frameworks must be established to ensure that the integration of AI aligns with the individual's wishes, respects personal autonomy, and addresses concerns related to identity preservation.

Moreover, responsible implementation and transparency are crucial when integrating AI into cryonics research and development. AI algorithms must be rigorously tested, validated, and regularly updated to ensure accuracy and reliability. Transparency in the use of AI can promote trust and facilitate critical evaluation of AI-assisted processes and decisions.

# The Role of AI in Future Revival Scenarios

Cryonics, the practice of preserving human bodies or brains at ultra-low temperatures with the hope of future revival, has long been associated with the vision of a future where cryopreserved individuals can be successfully revived. As advancements in artificial intelligence (AI) continue to unfold, the role of AI in future revival scenarios within the realm of cryonics becomes increasingly significant. Below we will explore the potential role of AI in future revival scenarios, examining how AI can assist in the restoration of cryopreserved individuals and the implications it holds for the field of cryonics.

One of the primary areas where AI can contribute to future revival scenarios is in the restoration of cognitive function and memory. Cryopreservation techniques aim to preserve the physical structures of the brain, but the revival process must go beyond mere preservation. AI algorithms can analyze the structural and functional data obtained from cryopreserved brains, simulate neural connections, and assist in the restoration of brain functionality. By utilizing AI in the reconstruction of damaged neural networks, scientists can potentially restore cognitive abilities and memories in cryopreserved individuals.

AI can also aid in the integration of cryopreserved individuals into future society. Revived individuals may face challenges in adapting to the technological, social, and cultural changes that have occurred during their period of cryopreservation. AI-based virtual reality simulations can provide immersive environments for cryopreserved individuals to acclimate and familiarize themselves with the new world they awaken to. These simulations can simulate real-life scenarios, offer training opportunities, and assist in the process of reintegration.

Moreover, AI can facilitate personalized medicine approaches in future revival scenarios. Each cryopreserved individual is unique, with distinct medical histories, genetic profiles, and treatment requirements. AI algorithms can analyze the vast amounts of data collected from the individual's cryopreservation and medical records to develop personalized treatment plans. By considering factors such as genetic predispositions, pre-existing conditions, and therapeutic responses, AI can assist in tailoring medical interventions specifically for each revived individual, optimizing their healthcare outcomes and enhancing their chances of long-term well-being.

Another potential role of AI in future revival scenarios is in the monitoring and management of revived individuals' health. Continuous health monitoring is vital for ensuring the well-being and longevity of cryopreserved individuals after revival. AI algorithms can analyze real-time data from wearable devices, medical sensors, and diagnostic tests to monitor various health parameters, detect early signs of disease or deterioration, and provide timely interventions. AI-based healthcare systems can offer personalized health recommendations, predict health risks, and facilitate proactive interventions, thereby optimizing the health outcomes of revived individuals.

Additionally, AI can contribute to ongoing research and development in the field of cryonics. AI algorithms can analyze large datasets of cryopreservation outcomes, tissue properties, and revival attempts to identify patterns, correlations, and factors that influence success rates. This knowledge can guide researchers in refining cryopreservation protocols, developing improved preservation techniques, and advancing the understanding of the revival process. By leveraging AI-driven insights, scientists can optimize the revival process and enhance the overall success of future revival scenarios.

However, the integration of AI in future revival scenarios also poses ethical and philosophical questions. Critics argue that the use of AI in the revival process may result in altered memories, personality, or subjective experiences, raising concerns about the preservation of personal identity. Ethical frameworks must be established to ensure that AI integration respects individual autonomy, aligns with the individual's wishes, and addresses identity-related concerns. Transparency and inclusiveness in decision-making processes are crucial to maintain the trust and well-being of revived individuals.

# Cryonics and Popular Culture

Cryonics, the practice of preserving human bodies or brains at ultra-low temperatures with the hope of future revival, has captured the imagination of people around the world. Its unique concept and potential implications for the future have not only fascinated scientists and enthusiasts but have also made their way into popular culture. Below we will explore the relationship between cryonics and popular culture, examining how cryonics has been depicted in various forms of media and the impact it has had on public perception.

One of the earliest instances of cryonics in popular culture can be traced back to science fiction literature. Authors such as H.G. Wells and Isaac Asimov have explored the concept of cryonics and immortality in their works. Wells' novel "The Sleeper Awakes" (1910) tells the story of a man who wakes up from a 200-year-long sleep to find himself in a future society. Asimov's "The Naked Sun" (1957) features a character who undergoes cryopreservation as a means of surviving long interstellar journeys. These early depictions of cryonics in literature laid the foundation for its subsequent appearances in popular culture.

Cryonics has also been prominently featured in film and television. One of the most iconic examples is the 1989 film "Bill & Ted's Excellent Adventure." In the movie, the protagonists travel through time, encountering various historical figures, including a cryonically preserved Napoleon Bonaparte. This comedic portrayal introduced cryonics to a wider audience, albeit in a lighthearted and fictional context. The film's popularity helped bring cryonics into the mainstream consciousness.

Another notable film featuring cryonics is the 1992 science fiction movie "Freejack." Set in a dystopian future, the film revolves around the concept of "freejacking," where the consciousness of a person about to die is transferred into the body of a living host. Cryonics plays a central role in the plot, with cryopreserved bodies being sought after by those seeking immortality. Although the film received mixed reviews, it further popularized the notion of cryonics and its potential implications for future technologies.

Cryonics has also made appearances in television shows, often exploring ethical and philosophical dilemmas associated with the practice. In the long-running series "Futurama," the main character, Philip J. Fry, accidentally falls into a cryogenic chamber and wakes up 1,000 years in the future. The show often touches upon themes of identity, mortality, and the consequences of being cryopreserved. "Black Mirror," a British anthology series known for its thought-provoking episodes, features an episode titled "San Junipero" where cryonics is explored in the context of a digital afterlife. These television portrayals of cryonics raise questions about the nature of consciousness, personal identity, and the potential impact of cryopreservation on individuals and society.

Cryonics has also found its way into popular music. The band "Megadeth" released a song titled "Countdown to Extinction" in 1992, which features the lyric "Cryonics, a future for us all." The song touches on the themes of mortality and the desire for eternal life through cryopreservation. This example demonstrates how cryonics has permeated popular culture beyond visual mediums and into the realm of music, allowing for further exploration and reflection on the topic.

The portrayal of cryonics in popular culture has undoubtedly influenced public perception and understanding of the practice. However, it is essential to recognize that these depictions often simplify the complexities of cryonics for dramatic effect or comedic purposes.

## Cryonics in Film, Television, and Literature

Cryonics, the practice of preserving human bodies or brains at ultra-low temperatures with the hope of future revival, has captivated the human imagination and has found its way into various forms of media, including film, television, and literature. The concept of cryonics has been explored, portrayed, and fictionalized in these mediums, often offering thought-provoking scenarios and raising ethical questions. Below we will delve into the representation of cryonics in film, television, and literature, examining its impact on public perception and the broader cultural discourse surrounding the subject.

Film has been one of the most prominent mediums for the portrayal of cryonics. In the 1980s, the film "Blade Runner" (1982), based on Philip K. Dick's novel "Do Androids Dream of Electric Sheep?", depicted a dystopian future where cryonics was used to extend the lives of humans and create replicants. The film raised philosophical questions about the nature of identity and the moral implications of artificially prolonging life. Another notable example is the 1997 film "Gattaca," set in a future where genetic engineering and cryonics play central roles. The film explores themes of discrimination, genetic determinism, and the quest for human perfection.

Television series have also engaged with the concept of cryonics. The animated show "Futurama" (1999-2013) revolves around a cryogenically preserved protagonist who wakes up in the distant future. The series humorously examines the challenges and possibilities of cryonics, touching on themes of personal identity, cultural dislocation, and the consequences of future technology. In the popular series "Black Mirror," the episode "San Junipero" explores cryonics in the context of a digital afterlife, questioning the nature of consciousness, the preservation of memories, and the ethical implications of achieving immortality through technology.

Literature has long been a source for exploring cryonics. H.G. Wells' novel "The Sleeper Awakes" (1910) tells the story of a man who wakes up in a future society after a prolonged sleep. The novel raises questions about the impact of societal and technological changes on the individual and the potential consequences of cryonic revival. Isaac Asimov's "The Naked Sun" (1957) features a protagonist who undergoes cryopreservation to survive long interstellar journeys, presenting cryonics as a means of achieving space exploration and colonization.

These representations of cryonics in film, television, and literature have had a significant impact on public perception and understanding of the practice. They have helped bring the concept of cryonics into the mainstream consciousness, allowing for broader discussions and explorations of its possibilities and implications. However, it is crucial to recognize that these depictions often simplify or fictionalize the science and process of cryonics for the sake of storytelling and dramatic effect. They should be approached with a critical eye and an understanding that they may not accurately reflect the current scientific understanding of cryonics.

The portrayal of cryonics in these mediums also raises important ethical and philosophical questions. The fictional scenarios often explore themes of identity, mortality, and the consequences of prolonging life through cryopreservation. They challenge us to consider the ethical implications of artificially extending life, the nature of personal identity in the face of long periods of suspended animation, and the potential impact of cryonics on society as a whole.

Additionally, the representation of cryonics in film, television, and literature has helped to humanize the practice and its participants. It has shed light on the hopes, fears, and aspirations of individuals who choose cryonics as a means of preserving their lives and identities.

# Public Perception of Cryonics

Cryonics, the practice of preserving human bodies or brains at ultra-low temperatures with the hope of future revival, has long been a subject of fascination and speculation. The concept of cryonics, with its promise of potentially extending life and overcoming mortality, evokes a range of responses and emotions from the public. Below we will explore the public perception of cryonics, examining the factors that shape public opinion, the common misconceptions surrounding the practice, and the implications for those involved in the field.

One factor influencing public perception of cryonics is the portrayal of the practice in popular media. Films, television shows, and literature often depict cryonics in sensationalized or speculative ways, which can contribute to misconceptions and misunderstandings. Cryonics is sometimes portrayed as a form of immortality or a magical solution to evade death. These portrayals can create unrealistic expectations and fuel skepticism or disbelief among the public. It is important to recognize that the depictions in popular media often simplify or fictionalize the science and process of cryonics for dramatic effect.

Another factor that shapes public perception of cryonics is the lack of widespread knowledge and understanding about the field. Cryonics involves complex scientific concepts and methodologies that may be unfamiliar to the general public. The technical nature of cryonics, including the process of vitrification and the challenges associated with revival, can be difficult to grasp without a background in biology or medicine. This lack of familiarity can lead to skepticism or a dismissive attitude towards cryonics as a fringe or pseudoscientific practice.

Additionally, ethical and philosophical concerns influence public perception of cryonics. Some individuals question the ethical implications of attempting to cheat death or view cryonics as an unjust allocation of resources, diverting attention and funding away from more immediate healthcare needs. Others raise philosophical questions about personal identity and the continuity of consciousness in cryonics. These concerns contribute to varying opinions within the public, with some viewing cryonics as a promising avenue for extending life and others dismissing it as scientifically dubious or ethically problematic.

Public perception of cryonics is also influenced by the credibility and transparency of cryonics organizations and practitioners. Trust is essential in an area that involves the preservation of one's body or brain for an indefinite period of time. Skepticism can arise if there is a lack of transparency, rigorous scientific research, or evidence supporting the claims made by cryonics organizations. Public perception can be shaped by the perception of cryonics as a legitimate scientific endeavor, backed by peer-reviewed research and conducted by reputable organizations.

Misconceptions surrounding cryonics further impact public perception. One common misconception is the idea that cryonics offers immortality or guarantees revival in the future. Cryonics proponents themselves acknowledge the speculative nature of revival and the uncertainties involved. Cryopreservation is not a guarantee of future revival, but rather a bet on the potential advancements of science and technology. Another misconception is that cryonics is only for the wealthy or elite. While the cost of cryopreservation can be significant, efforts are being made to explore more accessible and affordable options.

It is important to recognize that public perception of cryonics is diverse and evolving. Over the years, public interest and awareness of cryonics have increased, leading to a wider range of opinions and discussions. The scientific and medical communities have also engaged in debates and research surrounding cryonics, contributing to a more nuanced understanding of the field. As the field of cryonics continues to advance and more information becomes available, public perception may continue to evolve.

# Influential Cryonics Advocates in the Arts and Entertainment

Cryonics, the practice of preserving human bodies or brains at ultra-low temperatures with the hope of future revival, has captured the imagination of individuals across various fields, including arts and entertainment. Over the years, several influential figures in these industries have expressed their support for cryonics, bringing the practice into the spotlight and contributing to public discourse. Below we will explore some of the notable cryonics advocates in the arts and entertainment world, examining their contributions and the impact they have had on the perception and understanding of cryonics.

1. One of the most prominent advocates of cryonics in the entertainment industry is American author and futurist Ray Kurzweil. Known for his work in artificial intelligence and futurism, Kurzweil has been a vocal proponent of life extension technologies, including cryonics. He has written extensively on the subject, discussing the potential of cryonics in his books such as "The Age of Spiritual Machines" and "Fantastic Voyage: Live Long Enough to Live Forever." Kurzweil's advocacy has helped popularize the concept of cryonics and has contributed to a broader understanding of its potential implications.

Another influential figure in the arts who has publicly expressed support for cryonics is American science fiction writer Robert Ettinger. Often referred to as the "father of cryonics," Ettinger's groundbreaking book, "The Prospect of Immortality" (1964), introduced cryonics to a wider audience. His work laid the foundation for the modern cryonics movement, highlighting the scientific and philosophical aspects of the practice. Ettinger's advocacy and vision of cryonics as a means of overcoming mortality have inspired many individuals to explore the possibilities of cryonics.

In the music industry, the late American singer and songwriter David Bowie was known for his interest in cryonics. In a 1972 interview with Playboy magazine, Bowie expressed his fascination with the idea of cryonics and the potential for future revival. While his interest in cryonics was not explicitly reflected in his music, Bowie's public statements sparked discussions and curiosity surrounding the practice. His influence as a cultural icon helped draw attention to cryonics and its implications for the future of humanity.

In the world of film, Academy Award-winning filmmaker Woody Allen has featured cryonics in several of his movies. In the film "Sleeper" (1973), Allen portrays a man who is cryogenically frozen and wakes up 200 years later. The film uses cryonics as a comedic device, satirizing societal changes and exploring the clash between the past and the future. Allen's depiction of cryonics in "Sleeper" introduced the concept to a wider audience and highlighted its potential for storytelling and social commentary.

Furthermore, influential figures in the gaming industry have also embraced cryonics. American video game designer Warren Spector has expressed his interest in cryonics and its potential impact on future narratives and gaming experiences. Spector believes that cryonics can offer unique storytelling opportunities, allowing players to explore themes of identity, ethics, and the implications of extending human life. His advocacy highlights the creative and imaginative potential of cryonics in shaping future entertainment mediums.

The influence of these cryonics advocates in the arts and entertainment industry extends beyond their individual contributions. Their public support and exploration of cryonics have contributed to a broader cultural discourse surrounding the practice. By integrating cryonics into their work, these figures have sparked curiosity and prompted discussions about the potential of cryonics in extending human life, preserving personal identity, and shaping future narratives.

# Cryonics and Religion: Perspectives and Debates

Cryonics, the practice of preserving human bodies or brains at ultra-low temperatures with the hope of future revival, raises complex ethical and philosophical questions, including its compatibility with religious beliefs. The intersection of cryonics and religion has sparked debates among different faith traditions, as the practice challenges traditional notions of death, the afterlife, and the sanctity of the human body. Below we will explore the perspectives and debates surrounding cryonics within various religious contexts, examining how different religions approach the practice and the implications for their followers.

Christianity, one of the world's largest religions, encompasses a diverse range of beliefs and interpretations. Views on cryonics within Christianity vary widely, reflecting the theological and philosophical diversity within the faith. Some Christians argue that cryonics is incompatible with the belief in bodily resurrection, a central tenet of Christian faith. They argue that cryonics interferes with the natural process of death and resurrection as understood in Christian theology. Others, however, see cryonics as an extension of medical technology and a means to potentially preserve life until a future cure is discovered, aligning it with the Christian principle of valuing and preserving life.

In Islam, cryonics has received mixed reactions. Some Muslim scholars argue that cryonics is permissible as long as it does not contradict fundamental Islamic beliefs. They suggest that the practice can be seen as an attempt to save lives and alleviate suffering, which are considered noble endeavors in Islam. However, other scholars question the ethics of cryonics, expressing concerns about the interference with the natural process of death and the belief in the appointed time of one's death as decreed by God. The diversity of opinions within the Islamic faith reflects the ongoing theological discussions surrounding cryonics.

Buddhism, with its focus on impermanence and the cycle of birth and rebirth, offers an interesting perspective on cryonics. Some Buddhist scholars argue that cryonics aligns with the Buddhist principle of seeking liberation from suffering and extending the opportunity for enlightenment. They view cryonics as a form of compassionate action aimed at preserving life and the potential for spiritual growth. However, other Buddhists express reservations about cryonics, highlighting the impermanence of existence and the need to accept and embrace the natural process of death and rebirth.

Judaism, like other religions, contains a diversity of perspectives on cryonics. Orthodox Jewish views often emphasize the sanctity of the body and the importance of burial in accordance with Jewish law. They argue that cryonics interferes with the body's natural return to the earth and the process of spiritual elevation after death. However, some Jewish scholars in other denominations see cryonics as a valid scientific endeavor that aims to preserve life and alleviate suffering. They argue that cryonics can be compatible with Jewish values of pursuing knowledge and healing the world.

These are just a few examples of the perspectives and debates surrounding cryonics within religious contexts. It is important to recognize that religious beliefs are diverse and complex, and individuals within each faith tradition may hold different opinions on cryonics based on their personal interpretations and understanding of their religious teachings. The intersection of cryonics and religion invites ongoing dialogue and exploration of theological and ethical implications.

# Cryonics in the Context of Major World Religions

Cryonics, the practice of preserving human bodies or brains at ultra-low temperatures with the hope of future revival, raises profound questions in relation to major world religions. The intersection of cryonics and religious beliefs sparks debates and discussions about the nature of life, death, and the afterlife. Below we will explore how cryonics is viewed within the contexts of some of the world's major religions, examining their perspectives and implications.

Christianity, one of the largest religions globally, encompasses diverse interpretations and beliefs. Within Christianity, the views on cryonics vary. Some Christians argue that cryonics contradicts the teachings on the resurrection and the afterlife. They believe that life after death is a spiritual journey rather than a physical one, and cryonics interferes with the natural process of death and resurrection. However, other Christians see cryonics as an extension of medical technology, preserving life until a potential cure or revival is found. They view it as a way to value and respect life and explore the possibilities of future advancements in science and medicine.

Islam, another major world religion, offers different perspectives on cryonics. While there is no specific mention of cryonics in Islamic scripture, scholars have engaged in discussions about its permissibility. Some Islamic scholars argue that cryonics can be permissible as long as it does not contradict core Islamic beliefs. They view it as a means to preserve life and alleviate suffering, aligning with Islamic principles. However, other scholars express concerns about the interference with the natural process of death and the belief in the appointed time of one's death as decreed by God. These varying views reflect the ongoing theological debates within Islam.

Buddhism, known for its emphasis on impermanence and the cycle of birth and rebirth, offers an interesting perspective on cryonics. Buddhist teachings emphasize the acceptance of impermanence and the natural process of death and rebirth. Some Buddhist scholars argue that cryonics can be compatible with Buddhism if seen as an expression of compassion aimed at preserving life and the opportunity for spiritual growth. However, others question the ethics of cryonics, emphasizing the importance of accepting and embracing the impermanence of existence and the natural process of death.

In Judaism, cryonics has garnered various interpretations. Orthodox Judaism, which places a strong emphasis on the sanctity of the body, generally opposes cryonics. Orthodox Jewish views emphasize the importance of burial in accordance with Jewish law and the belief in bodily resurrection. Cryonics is seen as interfering with the body's return to the earth and the process of spiritual elevation after death. However, some Jewish scholars from other denominations view cryonics as a valid scientific endeavor that aims to preserve life and alleviate suffering. They argue that cryonics can be compatible with Jewish values of pursuing knowledge and healing the world.

It is important to note that these perspectives are not exhaustive, and within each religious tradition, there is a spectrum of views and interpretations. The intersection of cryonics and religion invites ongoing dialogue and exploration. Individual believers may hold diverse opinions on cryonics, based on their personal understanding and interpretations of their religious teachings.

It is also worth mentioning that while cryonics may raise theological questions within religious contexts, it is not inherently religious in nature. Cryonics is a scientific and medical practice that operates independently of religious beliefs. Individuals may choose cryonics for personal reasons, including their scientific outlook, desire for future possibilities, or personal views on life and death.

## Spiritual Implications of Cryonics

Cryonics, the practice of preserving human bodies or brains at ultra-low temperatures with the hope of future revival, raises profound spiritual questions and implications. The intersection of cryonics and spirituality explores the mysteries of life, death, and the nature of the human soul. Below we will delve into the spiritual implications of cryonics, examining its relationship with concepts such as the soul, consciousness, and the afterlife.

One of the key spiritual implications of cryonics is its impact on the understanding of the soul. Many spiritual traditions believe in the existence of a soul, a non-physical essence that persists beyond the physical body. Cryonics challenges the traditional notion of the soul by preserving the physical body while the soul is believed to have departed. This raises questions about the connection between the soul and the physical body, and whether the soul can be separated from the body and potentially return after cryopreservation.

The concept of consciousness is closely tied to the spiritual implications of cryonics. Consciousness is often considered the essence of individual identity and personal experience. Cryonics raises questions about the nature of consciousness and its potential preservation during the cryopreservation process. If consciousness is dependent on the physical brain, the preservation and revival of the brain may be seen as a way to maintain or restore consciousness in the future. However, if consciousness is considered independent of the physical body, cryonics may challenge the traditional understanding of consciousness and its relation to the soul.

Another spiritual consideration is the belief in the afterlife. Many spiritual traditions hold that death is a transition to another realm of existence, where the soul continues its journey or faces judgment. Cryonics introduces the possibility of interrupting this natural process of death and the journey of the soul. This raises questions about the impact of cryonics on the afterlife and whether the revival of cryopreserved individuals would disrupt or delay their spiritual journey.

The ethical and moral dimensions of cryonics also have spiritual implications. Some spiritual traditions emphasize the sanctity and naturalness of death, viewing it as part of the divine order. Cryonics challenges this perspective by intervening in the natural process of death and potentially extending life indefinitely. This raises ethical questions about the responsibility of individuals and the potential consequences of disrupting the natural cycle of life and death.

Additionally, the spiritual implications of cryonics extend to the emotional and psychological well-being of individuals. Death and grief are deeply spiritual experiences, often intertwined with questions of purpose, meaning, and the human condition. Cryonics offers hope and the possibility of continued existence, challenging the traditional understanding of mortality. This may provide solace to individuals grappling with the fear of death or seeking a sense of continuity beyond physical life.

It is important to note that the spiritual implications of cryonics are diverse and subjective, reflecting the beliefs and interpretations of individuals within different spiritual traditions. Views on cryonics can vary widely, even within the same tradition, as spiritual teachings and perspectives can be open to interpretation.

It is also crucial to recognize that the practice of cryonics itself is not inherently spiritual. Cryonics is a scientific and medical procedure aimed at preserving life and exploring the potential for future revival. Individuals may choose cryonics for a variety of reasons, including personal beliefs, scientific curiosity, or a desire to explore the possibilities of future advancements. Spiritual implications and considerations may inform the decision to pursue cryonics but do not necessarily define the practice itself.

# Reconciling Cryonics with Faith

Cryonics, the practice of preserving human bodies or brains at ultra-low temperatures with the hope of future revival, presents a unique challenge for individuals seeking to reconcile this scientific pursuit with their religious beliefs. The intersection of cryonics and faith requires careful consideration and exploration of theological perspectives. Below we will examine the ways in which individuals may reconcile cryonics with their faith, exploring different approaches and the implications for their spiritual journey.

One approach to reconciling cryonics with faith is to view it as an extension of the pursuit of knowledge and the alleviation of suffering. Many religious traditions emphasize the value of human life and the obligation to alleviate suffering. From this perspective, cryonics can be seen as a means to preserve life and potentially restore health in the future. By engaging in cryonics, individuals may perceive themselves as participating in scientific advancements and contributing to the well-being of humanity, aligning with their religious commitment to compassion and healing.

Another approach is to interpret cryonics as a form of hope and faith in the potential of future technologies. While religious beliefs often include concepts of an afterlife or spiritual realms, individuals may still hold onto the hope that scientific advancements, including cryonics, can offer possibilities for continued existence or future revival. This perspective allows individuals to integrate their religious faith with their curiosity about the potential of cryonics, embracing both spiritual beliefs and scientific exploration.

For some individuals, cryonics can be seen as an expression of trust in divine intervention or guidance. They may view cryonics as an opportunity for God to work through future scientific advancements, potentially enabling revival or the continuation of life. This perspective allows individuals to surrender the outcome of cryonics to a higher power, finding solace in the belief that the ultimate fate lies in the hands of their religious faith.

Additionally, individuals may approach cryonics as a personal choice based on their religious convictions and understanding of their faith's teachings. They may interpret their religious beliefs in a way that allows for the exploration of scientific possibilities, including cryonics, without conflicting with their core religious principles. This approach involves carefully examining the theological tenets and finding interpretations that are compatible with the pursuit of cryonics, ensuring that it does not violate religious values or beliefs.

However, it is important to acknowledge that reconciling cryonics with faith is not without challenges. Cryonics raises ethical questions and may challenge traditional views on death, the afterlife, and the sanctity of the human body. Different religious traditions have varying interpretations and teachings on these matters, and individual believers within a faith may have different perspectives. Reconciling cryonics with faith requires engaging in thoughtful reflection, studying religious texts, consulting with religious leaders, and engaging in dialogue with fellow believers.

Moreover, it is crucial to respect the diversity of religious beliefs and understand that individuals within the same faith tradition may hold different views on cryonics. The personal nature of faith means that each individual's reconciliation process will be unique and shaped by their specific religious beliefs, values, and interpretations.

# Cryonics and Animal Preservation

Cryonics, the practice of preserving human bodies or brains at ultra-low temperatures with the hope of future revival, has also found applications in the realm of animal preservation. While cryonics is primarily associated with the preservation of human life, the preservation of animals through cryonics has garnered attention and offers unique insights into the possibilities and challenges of the practice. Below we will explore the relationship between cryonics and animal preservation, examining the motivations, methods, and implications of preserving animal life through cryopreservation.

One of the primary motivations for animal preservation through cryonics is conservation efforts. Cryopreservation offers a means to preserve endangered or extinct animal species, potentially enabling their future revival or reintroduction into their natural habitats. By cryopreserving genetic material such as sperm, eggs, or embryos, scientists can store and protect the genetic diversity of endangered species, providing a potential lifeline for their survival. Cryonics thus plays a crucial role in safeguarding biodiversity and supporting conservation efforts.

Another aspect of animal preservation through cryonics is the study of biological processes and scientific research. Cryopreservation allows scientists to preserve animal tissues, organs, or cells at extremely low temperatures, effectively halting biological activity. This enables researchers to study these preserved samples in greater detail, providing insights into the structures, functions, and mechanisms of animal biology. Animal cryopreservation contributes to advancements in various fields, including medicine, veterinary science, and basic biological research.

In the realm of veterinary medicine, cryonics offers the possibility of preserving beloved companion animals. For pet owners who face the loss of their beloved furry friends, cryopreservation can provide a sense of hope and the potential for future revival. By preserving the animal's body or brain, pet owners envision the possibility of reuniting with their beloved companions in the future, perhaps when technological advancements or medical breakthroughs allow for their revival. Animal cryopreservation offers comfort and solace to grieving pet owners.

However, it is important to note that animal cryopreservation faces significant challenges and limitations. The complexities of cryopreservation vary among different animal species, and successful revival or restoration of preserved animals has not been achieved on a large scale. The process of cryopreservation and revival involves intricate biological processes and requires a deep understanding of the species-specific biology and physiological responses to extreme temperatures. Ethical considerations also come into play, as questions arise about the well-being and quality of life for revived animals.

Furthermore, the public perception of animal cryopreservation may differ from that of human cryopreservation. While cryopreservation of humans is primarily driven by personal beliefs, aspirations, and a desire for extended life, the cryopreservation of animals often focuses on conservation efforts, scientific research, and the preservation of genetic diversity. The motivation and ethical implications surrounding animal cryopreservation differ from those associated with human cryopreservation, underscoring the need for careful consideration and ethical guidelines.

# Applications of Cryopreservation for Endangered Species

Cryopreservation, the process of preserving biological materials at ultra-low temperatures, has emerged as a valuable tool for the conservation and protection of endangered species. By freezing and storing genetic material such as sperm, eggs, embryos, or tissues, cryopreservation offers a means to safeguard the genetic diversity of endangered species and potentially contribute to their future survival. Below we will explore the applications of cryopreservation for endangered species, examining its importance, methods, and implications in the context of wildlife conservation.

One of the key applications of cryopreservation for endangered species is the preservation of gametes, including sperm and eggs. By collecting and freezing sperm and eggs from endangered animals, conservationists can create a "genetic bank" that preserves the genetic material for future use. This technique, known as gamete cryobanking, provides a means to maintain genetic diversity and prevent the loss of valuable genetic traits within endangered populations. Gamete cryobanking is particularly useful when individuals cannot be physically brought together for breeding due to geographic, logistical, or behavioral constraints.

Cryopreservation also allows for the preservation of embryos, which offers additional advantages in the field of endangered species conservation. Embryo cryopreservation involves fertilizing eggs with cryopreserved sperm and then freezing and storing the resulting embryos. This method is particularly valuable for species with low fertility rates or those that are difficult to breed in captivity. By preserving embryos, scientists can potentially reintroduce them into the reproductive cycles of endangered animals or use them for artificial insemination, contributing to population growth and genetic diversity.

In addition to gametes and embryos, cryopreservation of tissues and cells has important applications in the conservation of endangered species. By collecting and preserving tissues, such as skin samples or blood samples, scientists can store valuable genetic information for future research and conservation efforts. Cryopreserved cells can be used for various purposes, including genetic analysis, cell culture, and the creation of induced pluripotent stem cells, which offer possibilities for regenerative medicine and cloning.

The application of cryopreservation for endangered species extends beyond preserving genetic material. It also enables the establishment of cryopreserved cell lines and genetic resources for research purposes. These resources serve as references for genetic studies, population analyses, and understanding the evolutionary history of endangered species. Cryopreserved genetic material can be used to investigate genetic variations, assess relatedness among populations, and inform conservation strategies aimed at preserving genetic diversity and promoting the long-term survival of endangered species.

Cryopreservation for endangered species also offers benefits in terms of wildlife management and assisted reproduction techniques. Cryopreserved genetic material can be used for artificial insemination, allowing the introduction of genetic diversity into small or isolated populations. It can also facilitate the creation of "insurance populations" or captive breeding programs, where cryopreserved genetic material can be used to reintroduce endangered species into their natural habitats when appropriate conditions are met.

However, it is essential to acknowledge the challenges and limitations of cryopreservation for endangered species. Cryopreservation techniques vary among different species, and successful revival or restoration of frozen genetic material has not been achieved uniformly. The success of cryopreservation depends on factors such as the type of tissue, the species-specific biology, and the availability of appropriate protocols. Moreover, the logistics of cryopreservation, including sample collection, storage, and transportation, require careful planning and adherence to scientific guidelines.

Ethical considerations also arise when applying cryopreservation techniques for endangered species. Questions may arise concerning the appropriate use of cryopreserved genetic material, the welfare of the animals involved, and the potential impact on the natural behavior and genetic adaptations of the species. These considerations necessitate the development of ethical guidelines and frameworks to ensure the responsible and sustainable application of cryopreservation techniques in endangered species.

# Companion Animal Cryonics

Cryonics, the practice of preserving human bodies or brains at ultra-low temperatures with the hope of future revival, has extended its reach beyond humans to include companion animals. Companion animal cryonics offers pet owners the possibility of preserving their beloved furry friends in the hope of a future revival. Below we will explore the concept of companion animal cryonics, examining the motivations, methods, and implications of this practice.

One of the primary motivations for companion animal cryonics is the emotional bond between pet owners and their beloved companions. Pets often become integral members of the family, providing love, companionship, and emotional support. When faced with the impending loss of a cherished pet, cryonics offers an opportunity for pet owners to extend their time together, albeit in a suspended state. The hope of future revival provides solace to grieving pet owners and offers the possibility of reuniting with their beloved companions.

The process of companion animal cryonics involves preserving the body or brain of the pet through cryopreservation techniques. After the pet passes away, the body or brain is carefully prepared and then cooled to extremely low temperatures. The aim is to halt biological decay and preserve the pet's physical structure and potential for future revival. The preserved remains are stored in specialized facilities where they are kept at cryogenic temperatures until such time as the technology and medical advancements necessary for revival become available.

While the concept of companion animal cryonics is intriguing to many pet owners, it is essential to understand the limitations and challenges associated with the practice. Cryopreservation is a complex process that requires advanced technology and expertise. Ensuring the successful preservation of a pet's body or brain necessitates meticulous attention to detail, from the rapid cooling of the remains to the careful handling and storage of the preserved specimen. However, it is crucial to acknowledge that the revival of cryopreserved animals has not yet been achieved on a large scale.

Moreover, ethical considerations arise in companion animal cryonics. Questions may arise concerning the welfare and quality of life for revived animals. It is essential to consider the physical and psychological well-being of animals if they were to be revived after cryopreservation. As the revival technology for cryopreserved animals is still hypothetical, the potential impact on their health, happiness, and overall welfare remains uncertain.

Public perception of companion animal cryonics is diverse, ranging from skepticism to hope. Some people see it as a way for pet owners to cling onto their loved ones and prolong the grieving process. Others view it as an expression of love and an opportunity to explore the potential of future advancements in science and technology. The ethical considerations and emotional attachments associated with companion animal cryonics require careful reflection and personal decision-making.

It is important to note that companion animal cryonics operates within a legal and regulatory framework. The availability of cryonics services for companion animals may vary depending on geographical location and local regulations. Pet owners considering companion animal cryonics should research and consult with reputable cryonics organizations to ensure compliance with legal requirements and to gain a thorough understanding of the process and its implications.

# Ethical Considerations for Animal Cryonics

Animal cryonics, the practice of preserving the bodies or genetic material of animals at ultra-low temperatures, raises important ethical considerations. As we explore the possibilities and implications of cryopreserving animals, it is crucial to address the ethical dimensions surrounding this practice. Below we will examine some of the key ethical considerations for animal cryonics, exploring topics such as animal welfare, the preservation of natural processes, and the allocation of resources.

One of the primary ethical considerations in animal cryonics is the welfare and well-being of the animals involved. Cryopreservation techniques and the subsequent revival process, if it becomes possible in the future, may raise questions about the physical and psychological well-being of the animals. It is essential to ensure that the animals experience minimal suffering during the cryopreservation process and, if revived, are provided with a good quality of life. Ethical guidelines and standards should be established to protect the welfare of the animals involved in cryopreservation and potential future revival.

Preserving the natural processes of life and death is another ethical consideration in animal cryonics. Some argue that cryopreservation disrupts the natural cycle of life and death, interfering with the process of decay and decomposition that is part of the natural order. Cryopreservation, from this perspective, may be seen as an unnatural intervention in the life cycle of animals. This raises questions about the inherent value of allowing animals to go through their natural life processes and whether cryopreservation is an appropriate interference.

Resource allocation is a significant ethical consideration in animal cryonics. The practice requires financial investment, infrastructure, and scientific resources. Critics argue that the resources used for cryopreservation and potential future revival could be better allocated to other pressing conservation efforts or animal welfare initiatives. Given the limited resources available, ethical questions arise about the prioritization of animal cryonics over other conservation or welfare projects that may have more immediate impact or address more pressing needs.

Another ethical concern is the potential impact of cryopreserved animals on natural ecosystems. If animals are revived and reintroduced into their natural habitats after cryopreservation, it is crucial to consider the ecological consequences. Revived animals may face challenges in adapting to their environments or interacting with other species. Careful assessment of the ecological impact and potential risks associated with reintroducing cryopreserved animals into their natural habitats is essential to avoid unintended harm to ecosystems.

Transparency and informed consent are ethical considerations that apply to both the collection of genetic material and the decision to cryopreserve animals. Proper consent should be obtained from individuals or organizations responsible for the animals, ensuring that the decision aligns with their values and intentions. In cases where animals are unable to provide consent, such as in the preservation of genetic material from deceased animals, ethical guidelines should be established to guide the decision-making process and protect the best interests of the animals.

Public perception and public engagement are also ethical considerations in animal cryonics. Transparency and communication with the public are essential to ensure that the practice is understood and accepted by society. Engaging in public dialogue and addressing concerns and ethical questions can contribute to the responsible and ethical development of animal cryonics. Educating the public about the benefits, limitations, and ethical implications of animal cryonics is crucial for fostering informed decision-making and fostering ethical practices.

# Cryonics and Space Exploration

The realm of cryonics, the practice of preserving human bodies or brains at ultra-low temperatures with the hope of future revival, intersects with the fascinating domain of space exploration. Cryonics offers potential opportunities and challenges in the context of space travel, colonization, and the long-duration missions required for interplanetary or even interstellar exploration. Below we will explore the relationship between cryonics and space exploration, examining the motivations, possibilities, and implications of using cryonics in the context of space travel.

One of the primary motivations for considering cryonics in space exploration is the challenge of long-duration space missions. Human space travel, particularly beyond Earth's orbit, presents significant challenges in terms of the duration of the journey, the effects of microgravity on the human body, and the limited resources available on spacecraft. Cryonics offers a potential solution by placing astronauts into a state of suspended animation, preserving their bodies during the journey and potentially reducing the resources needed for life support systems. By entering a cryopreserved state, astronauts could effectively "pause" their biological processes, conserving energy and minimizing the impact of prolonged space travel on their bodies.

Furthermore, cryonics could play a vital role in the colonization of other planets or celestial bodies. The journey to distant planets or star systems could take decades or even centuries using current propulsion technologies. Cryopreservation offers a means to transport humans over these immense distances, preserving their bodies and potentially reviving them upon arrival. Cryonics could serve as a bridge between generations, enabling the transfer of human life from one celestial body to another and ensuring the continuity of the human presence in space.

Cryonics also addresses the challenges associated with the limited resources available during space travel. The cryopreservation process could reduce the need for extensive food, water, and life support systems on spacecraft, conserving resources and extending the duration of missions. This has significant implications for deep space exploration, where resupply missions from Earth are not feasible. By reducing the resource requirements, cryonics could enable longer missions, broader scientific research, and more extensive exploration of our solar system and beyond.

However, it is important to acknowledge the limitations and challenges associated with cryonics in space exploration. The process of cryopreservation and subsequent revival is still hypothetical and has not been achieved on a large scale. The complex biological processes involved in cryopreservation, the potential damage to tissues during freezing and thawing, and the feasibility of successful revival pose significant scientific and technological challenges. Furthermore, the impact of long-term cryopreservation on human health, cognition, and aging is not yet fully understood.

Ethical considerations also arise when contemplating cryonics in space exploration. Questions may arise concerning the autonomy and consent of astronauts regarding the decision to undergo cryopreservation. Ensuring informed consent and respecting the individual's right to make choices regarding their own body are crucial ethical considerations. Additionally, the allocation of resources and the potential diversion of funding from other critical space exploration initiatives to cryonics research must be carefully examined.

Moreover, the public perception and acceptance of cryonics in space exploration are critical factors. Public understanding, support, and trust in the science behind cryonics are necessary for the integration of cryonics in space missions. Educating the public about the possibilities and challenges of cryonics in space exploration and engaging in public dialogue are essential for fostering informed decision-making and garnering support for this area of research.

# The Role of Cryonics in Long-Duration Space Missions

Cryonics, the practice of preserving human bodies or brains at ultra-low temperatures with the hope of future revival, has captured the imagination of scientists and space enthusiasts alike. As humanity looks toward the future of space exploration, the concept of cryonics emerges as a potential solution to the challenges of long-duration space missions. Below we will delve into the role of cryonics in long-duration space missions, exploring its motivations, possibilities, and implications.

One of the primary motivations for considering cryonics in long-duration space missions is the challenge of sustaining human life and well-being during extended periods in space. Traditional space missions, such as those to the International Space Station (ISS), typically last a few months. However, journeys to other celestial bodies, such as Mars, can take years due to the vast distances involved. Cryonics offers a potential solution by placing astronauts in a state of suspended animation, effectively pausing their biological processes and reducing the resources needed for life support systems during the journey.

The preservation of human bodies through cryonics could significantly mitigate the challenges posed by prolonged space travel. By lowering the body's temperature to a cryogenic state, the metabolic rate decreases, reducing the energy and resources required for life support systems. This approach allows astronauts to conserve vital resources such as food, water, and oxygen, thereby extending the mission's duration and reducing logistical constraints associated with resupply missions.

Another aspect of cryonics in long-duration space missions is the potential for space exploration and colonization. Cryopreservation could facilitate the transportation of humans to distant celestial bodies, such as Mars or beyond, where the journey could span decades or even centuries using current propulsion technologies. By preserving astronauts' bodies during the journey, cryonics could enable the continuity of human presence in space over vast time scales, allowing future generations to reach their destination and potentially be revived upon arrival.

However, it is important to acknowledge the limitations and challenges associated with cryonics in long-duration space missions. The process of cryopreservation and subsequent revival is still largely theoretical and has not been achieved on a large scale. The scientific and technological hurdles are considerable, as cryopreservation involves complex biological processes and the potential for tissue damage during freezing and thawing. Additionally, the long-term effects of cryopreservation on human health, cognition, and aging require further research and understanding.

Ethical considerations also arise when contemplating cryonics in long-duration space missions. Questions may arise regarding the autonomy and consent of astronauts in the decision to undergo cryopreservation. Ensuring informed consent and respecting an individual's right to make choices regarding their own body are crucial ethical considerations. Furthermore, the allocation of resources and the potential diversion of funding from other essential space exploration initiatives to cryonics research must be carefully evaluated.

Public perception and acceptance of cryonics in long-duration space missions also play a significant role. Public understanding, support, and trust in the science behind cryonics are essential for the integration of cryonics in space missions. Engaging in public dialogue, addressing concerns, and fostering informed decision-making are crucial steps toward garnering support for cryonics research in the context of long-duration space missions.

# Cryonics and Astrobiology: The Search for Extraterrestrial Life

Cryonics, the practice of preserving human bodies or brains at ultra-low temperatures with the hope of future revival, intersects with the fascinating field of astrobiology—the study of life in the universe. As scientists explore the possibility of finding extraterrestrial life, cryonics emerges as a relevant topic, offering insights and potential applications. Below we will explore the relationship between cryonics and astrobiology, examining the motivations, possibilities, and implications of cryonics in the search for extraterrestrial life.

One of the primary motivations for considering cryonics in astrobiology is the preservation of life forms that may exist beyond Earth. The search for extraterrestrial life involves exploring environments that are vastly different from our own. Some of these environments, such as the icy moons of Jupiter or Saturn, are characterized by extremely low temperatures. Cryonics offers a means to preserve and study potential extraterrestrial life forms, allowing scientists to maintain their viability and investigate their biology and potential for adaptation to extreme environments.

Cryopreservation can play a significant role in preserving samples collected from space missions. When exploring other celestial bodies, such as Mars or comets, scientists collect samples of soil, ice, or rock, which may contain traces of microbial life or organic molecules. Cryopreservation can ensure the long-term preservation of these samples, allowing for detailed analysis and potential revival if signs of life are discovered. Preserving these samples at cryogenic temperatures can safeguard the integrity of the organic materials and protect against degradation.

Additionally, cryonics may offer insights into the possibilities of panspermia—the hypothesis that life could be distributed throughout the universe via interstellar or interplanetary transport. If microbial life exists elsewhere in the universe, cryopreservation could enable the transport of these organisms to new environments, increasing the likelihood of colonization and the spread of life. By studying the preservation and revival of cryopreserved organisms, scientists can gain insights into the potential mechanisms of panspermia and the survival of life under extreme conditions.

However, it is important to acknowledge the challenges and limitations associated with cryonics in astrobiology. Cryopreservation techniques vary among different organisms, and successful revival or restoration of cryopreserved extraterrestrial life forms has not been achieved on a large scale. The complexities of cryopreservation, the potential damage to cellular structures during freezing and thawing, and the need to develop suitable revival techniques pose significant scientific and technological challenges.

Ethical considerations also arise when contemplating cryonics in astrobiology. Questions may arise regarding the preservation and treatment of potential extraterrestrial life forms. Respecting the autonomy and integrity of these organisms, ensuring that they are not exploited or subjected to unnecessary harm, and considering the potential ethical implications of studying or reviving extraterrestrial life forms require careful ethical reflection and consideration.

Furthermore, the public perception and acceptance of cryonics in the context of astrobiology are crucial factors. Public understanding and support for the science behind cryonics, as well as its potential applications in astrobiology, play a significant role in shaping the direction and funding of research in this field. Engaging in public dialogue, addressing concerns, and fostering informed decision-making are essential for promoting responsible and ethical practices in the intersection of cryonics and astrobiology.

# Technological Developments in Space Cryonics

Cryonics, the practice of preserving human bodies or brains at ultra-low temperatures with the hope of future revival, has captured the imagination of scientists and space enthusiasts alike. As humanity continues to push the boundaries of space exploration, technological advancements in cryonics become increasingly relevant. Below we will explore the technological developments in space cryonics, examining the motivations, progress, and implications of these advancements.

One of the primary motivations for technological developments in space cryonics is the potential to extend human life and enable long-duration space missions. Traditional space missions, such as those to the International Space Station (ISS), typically last for months. However, future missions to distant celestial bodies, such as Mars or beyond, will require astronauts to spend extended periods in space, potentially lasting years or even decades. Cryonics offers a potential solution by placing astronauts in a state of suspended animation, effectively pausing their biological processes and reducing the resources needed for life support systems during these long-duration missions.

Technological advancements in cryopreservation techniques play a crucial role in making space cryonics a reality. Cryopreservation involves cooling the body to ultra-low temperatures while minimizing the formation of ice crystals that can damage tissues. New methods and technologies are being developed to improve the cryopreservation process, such as the use of cryoprotectants to protect cells and tissues from freezing damage, and the development of specialized cooling techniques that allow for rapid and uniform cooling. These advancements aim to improve the viability of cryopreserved organisms and increase the chances of successful revival in the future.

Another technological development in space cryonics relates to the storage and transportation of cryopreserved bodies or genetic material. Cryopreservation requires the use of specialized facilities to maintain ultra-low temperatures, ensuring the long-term preservation of the specimens. Technological advancements in cryogenic storage, including improvements in insulation materials and cooling systems, contribute to more efficient and reliable storage capabilities for long-duration space missions. Additionally, advancements in cryogenic transportation, such as the development of advanced cryogenic containers and the optimization of thermal management systems, enable the safe and efficient transport of cryopreserved specimens during space missions.

Furthermore, progress in revival technologies is a critical aspect of technological developments in space cryonics. While the revival of cryopreserved organisms or humans remains a theoretical possibility, ongoing research focuses on understanding the challenges and developing potential solutions. Nanotechnology, for instance, has been proposed as a potential tool for repairing cellular and molecular damage that occurs during the cryopreservation process. By developing nanoscale repair mechanisms, scientists aim to restore the structure and functionality of the cryopreserved organisms upon revival.

However, it is essential to acknowledge the limitations and challenges associated with technological developments in space cryonics. Cryopreservation and revival technologies are complex and still largely theoretical, with many scientific and technical hurdles to overcome. The potential damage to tissues during freezing and thawing, the impact of cryopreservation on cellular structures, and the effects of prolonged cryopreservation on overall viability and health remain significant challenges that require further research and development.

Ethical considerations also arise when contemplating technological developments in space cryonics. Questions may arise regarding the autonomy and consent of individuals regarding the decision to undergo cryopreservation and potential revival. Ensuring informed consent and addressing potential ethical concerns, such as the potential impact on personal identity or the welfare of revived individuals, require careful consideration and the development of ethical guidelines and frameworks.

Public perception and acceptance of technological developments in space cryonics are crucial factors. Public understanding and support for the science behind cryonics, as well as its potential applications in space exploration, play a significant role in shaping the direction and funding of research in this field.

# Cryonics and Cryobiology

Cryonics, the practice of preserving human bodies or brains at ultra-low temperatures with the hope of future revival, shares a close relationship with the field of cryobiology. Cryobiology, the study of the effects of low temperatures on biological systems, provides the scientific foundation for cryonics. Below we will explore the connection between cryonics and cryobiology, examining the motivations, principles, and implications of their interplay.

One of the primary motivations for the intersection of cryonics and cryobiology is the understanding of how living organisms respond to extreme cold temperatures. Cryonics relies on the principles of cryobiology to ensure the successful preservation and potential revival of cryopreserved bodies or brains. Cryobiologists study the mechanisms by which cells and tissues can be preserved at low temperatures, with the goal of minimizing the damage caused by freezing and thawing processes.

Cryobiology provides valuable insights into the cryopreservation process, including the optimization of cryoprotectants. Cryoprotectants are substances used to protect biological materials from freezing damage by reducing the formation of ice crystals. Cryobiologists study the properties of different cryoprotectants, their effects on cell viability, and their ability to prevent ice formation. These insights guide cryonics researchers in selecting and developing cryoprotectants that are effective in preserving the structural and functional integrity of the preserved tissues.

Furthermore, cryobiology investigates the impact of low temperatures on cellular and molecular processes. The freezing and thawing processes during cryopreservation can cause cellular damage due to the formation of ice crystals, osmotic stress, and intracellular ice formation. Cryobiologists study the mechanisms of cell injury and develop strategies to minimize damage, such as optimizing cooling rates and developing techniques for controlled thawing. This knowledge is vital in cryonics to mitigate the potential harm caused by cryopreservation and maximize the chances of successful revival in the future.

Cryobiology also explores the mechanisms by which certain organisms naturally withstand extreme cold temperatures. Some organisms, such as certain species of plants, insects, or microorganisms, have evolved mechanisms to survive freezing temperatures. Cryobiologists study these organisms to understand the physiological and biochemical adaptations that allow them to tolerate extreme cold. These adaptations can inform the development of novel approaches and technologies for cryopreservation in cryonics.

However, it is important to acknowledge the challenges and limitations associated with the application of cryobiology principles to cryonics. Cryobiology research primarily focuses on the preservation of cells, tissues, or small organisms, while cryonics aims to preserve whole human bodies or brains. The scale and complexity of cryonics present additional challenges, such as the need to preserve and revive complex organ systems and the potential damage caused by long-term storage at ultra-low temperatures.

Ethical considerations arise when discussing the application of cryobiology in cryonics. Cryonics involves the preservation of deceased individuals with the hope of future revival, raising questions about the ethical implications of such practices. Considerations such as consent, respect for human remains, and the allocation of resources for cryonics research versus other critical areas of healthcare and scientific advancement require careful ethical reflection.

Public perception and acceptance of cryobiology and its application in cryonics are important factors. Understanding and support for the scientific principles and technological advancements in cryobiology help foster public trust in the cryonics field. Communicating the scientific underpinnings of cryonics, addressing concerns, and fostering informed discussions are vital for public engagement and acceptance.

# Cryopreservation of Tissues and Organs

Cryopreservation, the process of preserving tissues and organs at ultra-low temperatures, plays a vital role in various fields, including medicine, transplantation, and research. The preservation of tissues and organs through cryopreservation offers numerous possibilities and applications, with potential implications for the advancement of healthcare and scientific understanding. Below we will explore the fascinating world of cryopreservation of tissues and organs, examining its motivations, methods, and implications in relation to cryonics.

One of the primary motivations for the cryopreservation of tissues and organs is the preservation of biological materials for medical purposes. Cryopreservation enables the storage of tissues and organs for extended periods, allowing them to be used for various medical interventions and therapies. For instance, cryopreserved tissues, such as corneas, can be used for transplantation, restoring vision for individuals with corneal damage. Cryopreservation also facilitates the storage of organs for transplantation, extending the window of time for matching recipients and increasing the availability of organs for those in need.

The methods employed in the cryopreservation of tissues and organs vary depending on the specific material being preserved. One common approach involves the use of cryoprotectants, substances that protect the cells and tissues from damage during freezing and thawing. Cryoprotectants can penetrate the cells, replacing the water molecules and reducing the formation of ice crystals, which can cause cellular damage. Gradual cooling and controlled freezing techniques are employed to minimize cellular damage during the freezing process. Preservation at ultra-low temperatures, typically below -130 degrees Celsius (-202 degrees Fahrenheit), helps to maintain the long-term viability of the cryopreserved tissues and organs.

Cryopreservation has revolutionized the field of organ transplantation. Organs that have been cryopreserved can be stored for longer periods, enabling more efficient organ allocation and transplantation. Cryopreservation also allows for the creation of organ banks, where organs can be stored until a suitable recipient is available. This extends the lifespan of organs and improves the chances of successful transplantation.

Moreover, cryopreservation of tissues and organs has significant implications in the field of regenerative medicine. The preservation of stem cells and other specialized cells at low temperatures allows for their later use in tissue engineering and regenerative therapies. Cryopreservation enables the storage of cells with specific characteristics, ensuring a constant supply of viable cells for research and therapeutic purposes. This opens up possibilities for the development of advanced treatments for various medical conditions, including the regeneration of damaged tissues and the creation of artificial organs.

However, challenges and limitations exist in the cryopreservation of tissues and organs. Cryopreservation can cause cellular damage due to the formation of ice crystals, osmotic stress, and intracellular ice formation. These factors can impact the viability and functionality of the preserved tissues and organs. Research is ongoing to optimize cryopreservation techniques and develop strategies to minimize damage and enhance the success rates of preservation and subsequent revival.

Ethical considerations also arise in the cryopreservation of tissues and organs. Questions may arise concerning the allocation of resources and access to cryopreserved tissues and organs. Ensuring equitable distribution, fair allocation policies, and addressing potential ethical concerns related to consent and the use of cryopreserved materials are crucial aspects of responsible and ethical cryopreservation practices.

Public perception and acceptance of cryopreservation of tissues and organs are also significant factors. Educating the public about the scientific principles and potential benefits of cryopreservation helps foster understanding and support for this field of research. Public engagement, addressing concerns, and promoting transparency are essential for gaining public trust and support.

# The Science of Cryogenics and Low-Temperature Biology

Cryogenics, the branch of science that deals with the production and effects of extremely low temperatures, plays a pivotal role in the field of cryonics. Cryonics, the practice of preserving human bodies or brains at ultra-low temperatures with the hope of future revival, relies on the principles and advancements in cryogenics and low-temperature biology. Below we will explore the fascinating science of cryogenics and low-temperature biology, examining the motivations, principles, and implications of these fields in relation to cryonics.

One of the primary motivations for studying cryogenics and low-temperature biology is the understanding of how living organisms and materials respond to extreme cold temperatures. The science of cryogenics delves into the properties of materials and the physical and chemical changes that occur at low temperatures. By studying the behavior of matter at these frigid temperatures, scientists gain insights into the potential effects on living organisms and develop techniques to mitigate the damage caused by extreme cold.

Cryogenics encompasses the production and use of cryogenic substances, such as liquid nitrogen and helium, which reach extremely low temperatures. These substances have various applications, including preserving biological materials, conducting experiments, and facilitating superconductivity. Liquid nitrogen, in particular, is commonly used in cryonics to achieve the ultra-low temperatures required for preserving bodies or brains.

Low-temperature biology, on the other hand, focuses specifically on the effects of low temperatures on biological systems. It investigates the physiological and biochemical changes that occur in organisms when exposed to cold environments. Low-temperature biology explores how organisms adapt to survive and even thrive in frigid conditions, such as the Arctic or Antarctic regions. This field also explores the mechanisms by which certain organisms, such as polar bears or certain plants, are capable of withstanding extreme cold temperatures.

The science of cryogenics and low-temperature biology provides crucial insights into the cryopreservation process employed in cryonics. Cryopreservation involves cooling biological materials to extremely low temperatures to slow down or halt the processes of decay and degradation. By reducing the temperature, cellular activities, including metabolism, slow down, which helps to preserve the structural integrity and biochemical composition of the preserved material.

To achieve successful cryopreservation, cryobiologists employ various techniques and methodologies. Cryoprotectants, substances that protect cells and tissues from damage during freezing and thawing, are often used. These cryoprotectants help prevent the formation of ice crystals, which can be detrimental to the cellular structure. Moreover, cooling and freezing rates, as well as the controlled thawing process, are carefully managed to minimize cellular damage and maximize the chances of successful revival in the future.

However, it is essential to acknowledge the challenges and limitations associated with the science of cryogenics and low-temperature biology in cryonics. Cryopreservation techniques and the subsequent revival process are still largely experimental and face scientific and technical hurdles. The potential damage to tissues during freezing and thawing, the optimization of cryoprotectant formulations, and the long-term effects of cryopreservation on cellular structures and functionality require further research and development.

Ethical considerations also arise when discussing the science of cryogenics and low-temperature biology in relation to cryonics. Questions may arise concerning consent and the implications of preserving human bodies or brains for potential revival in the future. Respecting the autonomy and wishes of individuals regarding their own bodies, addressing concerns about personal identity, and developing ethical guidelines and frameworks are crucial aspects of responsible cryonics practices.

# Applications of Cryobiology in Medicine and Research

Cryobiology, the study of the effects of low temperatures on biological systems, has found a wide range of applications in the fields of medicine and research. The principles and techniques developed in cryobiology have revolutionized various areas, including organ transplantation, fertility preservation, and the preservation of biological samples. Below we will explore the fascinating applications of cryobiology in medicine and research, and their relation to cryonics.

One of the primary applications of cryobiology is in the field of organ transplantation. Cryopreservation techniques enable the long-term storage of organs, extending their viability beyond the typical time constraints. By cooling organs to ultra-low temperatures, cryobiologists can preserve their structural and functional integrity, allowing for more efficient organ allocation and matching with suitable recipients. Cryobiology has significantly improved the success rates of organ transplantation and has increased the availability of organs for patients in need.

Cryobiology also plays a crucial role in the preservation of reproductive tissues and cells. In the field of fertility preservation, cryopreservation techniques are employed to store eggs, sperm, and embryos. This allows individuals who are undergoing medical treatments that may impact their fertility, such as chemotherapy or radiation, to preserve their reproductive potential for future use. Cryopreservation of reproductive tissues has also paved the way for assisted reproductive technologies, such as in vitro fertilization (IVF), by providing a means to store and preserve embryos.

Furthermore, cryobiology has advanced the field of regenerative medicine. The preservation of stem cells and other specialized cells at ultra-low temperatures allows for their long-term storage and later use in tissue engineering and regenerative therapies. Cryopreservation ensures a constant supply of viable cells, enabling researchers and clinicians to develop advanced treatments for various medical conditions. The use of cryopreserved cells in regenerative medicine has shown promise in tissue repair, organ regeneration, and the development of personalized therapies.

Cryobiology also plays a significant role in the preservation and storage of biological samples for research purposes. Cryopreservation techniques allow for the long-term storage of cells, tissues, and genetic material, preserving their viability and integrity. This has profound implications in areas such as biobanking, genetic research, and the study of rare or endangered species. Cryopreserved samples serve as valuable resources for future investigations, facilitating scientific breakthroughs and advancements in various fields of research.

Moreover, cryobiology has found applications in the field of cryosurgery. Cryosurgery involves the use of extreme cold temperatures to treat various conditions, including certain types of cancers and skin lesions. By applying cryogenic substances, such as liquid nitrogen, to the targeted area, cryosurgeons can selectively freeze and destroy abnormal or diseased cells. Cryosurgery offers a minimally invasive approach with fewer side effects compared to traditional surgical techniques.

While cryobiology has made significant contributions to medicine and research, there are challenges and limitations associated with its applications. Cryopreservation techniques are not without risks, and the success rates can vary depending on the specific material being preserved. The formation of ice crystals during freezing and thawing processes can potentially damage cells and tissues, requiring further advancements in cryoprotectants and preservation methods to minimize these risks.

Ethical considerations also arise when discussing the applications of cryobiology in medicine and research. Questions may arise concerning the allocation of resources and access to cryopreserved materials, as well as the implications of cryopreserving human bodies or brains for potential revival in the future. Ensuring equitable distribution, fair allocation policies, and addressing potential ethical concerns related to consent and the use of cryopreserved materials are crucial aspects of responsible and ethical practices in cryobiology.

# The Revival Process: Reanimation and Reintegration

Cryonics, the practice of preserving human bodies or brains at ultra-low temperatures with the hope of future revival, raises fascinating questions about the process of reanimating and reintegrating cryopreserved individuals into society. While the revival process remains speculative and hypothetical, the concept of reanimation and reintegration is a topic of great interest and speculation. Below we will explore the intriguing possibilities, challenges, and implications associated with the revival process in relation to cryonics.

The process of reanimating and reintegrating cryopreserved individuals into society raises numerous intriguing questions. One of the primary considerations is the revival of cognitive functions and consciousness. Cryonics aims to preserve the structure and integrity of the brain, including its neural connections and stored memories. The challenge lies in the revival process itself, where scientists would need to address the restoration of neural activity and the revival of individual consciousness. The understanding of brain function, memory, and consciousness remains an active area of research, and the revival process would require significant advancements in neuroscience and neurotechnology.

Another aspect of reanimation and reintegration is the physical rehabilitation and adaptation of cryopreserved individuals to the modern world. If successful, cryonics could revive individuals from different time periods, potentially leading to a significant cultural and societal shift. The reintroduction of cryopreserved individuals would require physical rehabilitation, medical care, and psychological support to help them adjust to the changes in society, technology, and social norms. This process would necessitate extensive interdisciplinary collaboration between medical professionals, psychologists, sociologists, and ethicists.

Additionally, the revival process raises ethical considerations. Questions may arise regarding the rights and identity of revived individuals, as well as their integration into society. Respecting the autonomy and wishes of cryopreserved individuals, ensuring informed consent, and addressing potential ethical concerns related to personal identity, societal impact, and the allocation of resources require careful consideration. Ethical frameworks and guidelines would need to be developed to ensure the responsible and ethical treatment of revived individuals and the fair distribution of resources during the revival and reintegration process.

The revival and reintegration of cryopreserved individuals also pose logistical challenges. Cryonics facilities would need to establish protocols and systems for the revival process, including the provision of medical care, rehabilitation, and support services. This would involve the collaboration of experts in various fields, including cryobiology, medicine, psychology, and social sciences. Adequate infrastructure and resources would be required to accommodate the potential influx of revived individuals and to ensure their well-being and successful reintegration into society.

Furthermore, public perception and acceptance play a significant role in the revival and reintegration of cryopreserved individuals. Public understanding and support for the science behind cryonics, as well as the potential implications of revival, are crucial for the successful reintegration of revived individuals into society. Education, public dialogue, and informed decision-making are vital for fostering public acceptance and addressing concerns surrounding the revival and reintegration process.

It is essential to note that the revival process in cryonics is largely speculative and faces significant scientific, technological, ethical, and societal challenges. The field of cryonics is still in its early stages, and the scientific feasibility and ethical implications of revival remain areas of ongoing research and debate. It is important to approach discussions on the revival process with scientific skepticism and careful consideration of the complex factors involved.

## The Science of Reanimation

Cryonics, the practice of preserving human bodies or brains at ultra-low temperatures with the hope of future revival, raises intriguing questions about the scientific principles and processes involved in the hypothetical scenario of reanimation. While the concept of reanimation remains speculative and hypothetical, the scientific aspects surrounding the potential revival of cryopreserved individuals are a subject of great interest and curiosity. Below we will explore the science of reanimation in relation to cryonics, examining the motivations, challenges, and implications associated with this captivating concept.

The science of reanimation primarily revolves around the revival of cognitive functions and consciousness. Cryonics aims to preserve the structure and integrity of the brain, including its neural connections and stored memories. However, the revival process presents numerous challenges. One of the key considerations is the restoration of neural activity, which is crucial for the revival of cognitive functions. Understanding the complexities of neural circuits, synaptic plasticity, and information processing in the brain is essential to develop the scientific framework for reanimation.

Advancements in neuroscience and neurotechnology play a vital role in the science of reanimation. Scientists are continually striving to understand the mechanisms underlying brain function, memory formation, and consciousness. Techniques such as optogenetics, which use light to manipulate neural activity, and brain-computer interfaces, which establish communication between the brain and external devices, hold promise for the development of potential reanimation strategies. These advancements in neuroscience provide valuable insights into the restoration of neural function and pave the way for potential breakthroughs in reanimation.

Another aspect of the science of reanimation involves addressing the challenges associated with the preservation and restoration of memories. Memories are intricately linked to the structure and function of the brain. The successful revival of cryopreserved individuals would require methods to restore and integrate preserved memories into the reanimated brain. Understanding the neural processes underlying memory formation, storage, and retrieval is crucial for the development of techniques that can potentially restore memories in revived individuals.

Additionally, the science of reanimation explores the physiological and biochemical challenges associated with the revival process. Cryopreservation involves lowering the temperature to ultra-low levels, which can cause cellular damage. The revival process would require techniques to mitigate this damage and restore cellular function. Advances in nanotechnology and regenerative medicine hold promise for repairing cellular and molecular damage that occurs during cryopreservation, potentially enabling the successful reanimation of cryopreserved individuals.

Moreover, the restoration of bodily functions and physical rehabilitation are essential aspects of reanimation. Cryopreserved individuals would need to regain motor control, sensory perception, and overall physical well-being. Rehabilitation therapies, such as physiotherapy and occupational therapy, would be crucial in assisting revived individuals in regaining their physical abilities and adapting to the changes that occur during the cryopreservation process.

Ethical considerations arise when discussing the science of reanimation. Questions may arise regarding the rights, identity, and autonomy of revived individuals. Respecting the wishes and autonomy of cryopreserved individuals, ensuring informed consent, and addressing potential ethical concerns related to personal identity, societal impact, and the allocation of resources require careful consideration. Ethical frameworks and guidelines would need to be developed to ensure responsible and ethical practices during the reanimation process.

It is important to note that the science of reanimation is highly speculative and faces significant scientific, technological, ethical, and societal challenges. The field of cryonics is still in its early stages, and the scientific feasibility and ethical implications of reanimation remain areas of ongoing research and debate. The complexities of brain function, memory restoration, and cellular repair pose substantial scientific hurdles that require further exploration and advancement.

# Psychological and Societal Implications of Revival

Cryonics, the practice of preserving human bodies or brains at ultra-low temperatures with the hope of future revival, raises thought-provoking questions about the psychological and societal implications of potential reanimation. While the concept of revival remains speculative and hypothetical, the idea of bringing cryopreserved individuals back to life has profound psychological and societal ramifications. Below we will explore the psychological and societal implications of revival in relation to cryonics, examining the motivations, challenges, and potential impact on individuals and society.

The psychological implications of revival encompass various aspects, including personal identity, psychological well-being, and existential considerations. Reviving individuals from cryopreservation would confront them with the challenge of adapting to a world that has evolved during their absence. The psychological adjustment to a changed society, with advancements in technology, cultural shifts, and new social norms, may pose significant challenges for revived individuals. This adjustment process would require support, counseling, and assistance to navigate the complexities of reintegrating into a society that may be vastly different from the one they left.

Personal identity is another crucial psychological consideration. Revived individuals would need to grapple with questions about their identity, self-concept, and place in the world. The passage of time, changes in personal relationships, and the potential loss of connections to the past may affect their sense of self. Psychological support and therapy would be essential in helping individuals explore and reconcile their past and present identities, facilitating a sense of continuity and integration.

The revival process also raises existential questions and challenges individuals' beliefs and perspectives. Revived individuals would need to confront questions about the nature of life, death, and the meaning of their revived existence. This existential journey may require philosophical contemplation, spiritual exploration, and support to find meaning and purpose in their renewed lives.

Societal implications of revival extend beyond the individual level and encompass broader considerations. The reintroduction of cryopreserved individuals into society would have social, cultural, and economic consequences. The arrival of revived individuals from different time periods may challenge established societal norms and practices. The blending of diverse perspectives, beliefs, and values could lead to significant cultural shifts and social transformation. Society would need to adapt to accommodate the unique needs, perspectives, and contributions of revived individuals while ensuring social cohesion and harmony.

The reintegration of revived individuals would necessitate support systems and policies to facilitate their successful integration into society. Psychosocial services, education, and vocational training would be crucial in helping revived individuals acquire the necessary skills, knowledge, and competencies to thrive in the modern world. Addressing potential social stigmatization, promoting inclusivity, and fostering a sense of belonging are important considerations in creating an environment conducive to the well-being and integration of revived individuals.

Ethical considerations arise when discussing the psychological and societal implications of revival. Questions may arise regarding the allocation of resources, access to revival technologies, and the potential impact on society. Ensuring equitable distribution, fair allocation policies, and addressing potential ethical concerns related to consent, social equality, and resource allocation require careful consideration and the development of ethical frameworks and guidelines.

Public perception and acceptance of revival also play a significant role in shaping its psychological and societal implications. Public understanding and support for the science behind cryonics, as well as the potential societal impact of revival, are crucial for fostering social acceptance and integration. Education, public dialogue, and informed decision-making are essential for addressing concerns, dispelling misconceptions, and promoting an inclusive and supportive environment for revived individuals.

# Future Technologies for Reintegration and Adaptation

Cryonics, the practice of preserving human bodies or brains at ultra-low temperatures with the hope of future revival, sparks curiosity about the potential technologies that could aid in the reintegration and adaptation of cryopreserved individuals. While the idea of revival remains speculative, exploring future technologies that could assist in the reintegration process opens up exciting possibilities. Below we will delve into the realm of future technologies for reintegration and adaptation, examining their motivations, potential advancements, and implications in relation to cryonics.

One of the primary motivations for developing future technologies for reintegration and adaptation is to assist cryopreserved individuals in navigating the changes that occur during their absence. The passage of time may bring about significant advancements in technology, communication, and societal structures. Future technologies could bridge the gap between the past and present, enabling cryopreserved individuals to acclimate to the modern world more smoothly.

Advancements in virtual reality (VR) and augmented reality (AR) hold promise for facilitating the reintegration process. VR and AR technologies could recreate historical environments, allowing revived individuals to immerse themselves in the familiar settings of their past. By simulating past eras, individuals could reacquaint themselves with societal norms, cultural practices, and personal memories. These immersive experiences could aid in psychological adjustment, memory restoration, and the development of a sense of continuity.

Artificial intelligence (AI) and machine learning (ML) technologies also offer significant potential for reintegration and adaptation. AI algorithms could assist revived individuals in understanding and navigating the modern world. AI-powered personal assistants could provide information, guidance, and support tailored to the unique needs and challenges of each individual. Natural language processing capabilities would facilitate communication and enable personalized interactions, fostering a sense of connection and belonging.

Furthermore, advancements in robotics could enhance physical rehabilitation and assist in daily tasks. Robotic exoskeletons and prosthetic limbs could restore mobility and motor functions, aiding cryopreserved individuals in adapting to changes in physical abilities. Humanoid robots could provide companionship and support, easing the potential sense of isolation and loneliness that may arise during the reintegration process.

Biotechnological advancements also have potential implications for reintegration and adaptation. Gene editing technologies, such as CRISPR-Cas9, offer the possibility of genetic modifications to address health issues and optimize physiological functions. Genetic therapies could be utilized to reverse aging-related changes, improve immune system function, and enhance overall well-being. These advancements would contribute to the physical well-being and longevity of revived individuals, supporting their successful reintegration into society.

Communication technologies are instrumental in facilitating connections and social integration. Future developments in telecommunication, internet connectivity, and social networking platforms could aid revived individuals in connecting with their loved ones, building new relationships, and participating in social networks. Virtual communities and online support groups specific to revived individuals could provide a sense of belonging and understanding, fostering social integration and emotional well-being.

Ethical considerations arise when discussing future technologies for reintegration and adaptation. Questions may arise regarding equitable access to these technologies, potential risks, and unintended consequences. Ensuring equal opportunities and accessibility, addressing potential privacy concerns, and safeguarding against misuse of emerging technologies require careful ethical considerations and regulations.

Public perception and acceptance of future technologies for reintegration and adaptation are crucial factors. Public understanding and support for the science behind cryonics and the potential benefits of these technologies are vital for fostering research, development, and implementation. Public engagement, education, and open dialogue are essential for addressing concerns, dispelling misconceptions, and ensuring ethical and responsible use of future technologies.

# Cryonics Case Studies: Real-Life Stories and Experiences

Cryonics, the practice of preserving human bodies or brains at ultra-low temperatures with the hope of future revival, has captured the imagination of many individuals seeking a chance at extended life or even immortality. While the scientific feasibility and ethical implications of cryonics remain subjects of ongoing debate, there are real-life case studies that shed light on the motivations, experiences, and outcomes of those who have chosen cryopreservation. Below we will explore some notable cryonics case studies, examining their stories and the impact they have had on the field of cryonics.

One of the most prominent cryonics case studies is that of Dr. James Bedford. In 1967, Dr. Bedford became the first person to be cryopreserved. His decision to undergo cryonics was motivated by his belief in the potential of future medical advancements to revive him. Dr. Bedford's body was preserved in liquid nitrogen and has remained cryopreserved to this day. His case sparked public interest and marked the beginning of cryonics as a recognized practice.

Another notable case is that of Dora Kent. In 1987, Dora Kent, a prominent figure in the cryonics community, was cryopreserved following her death. However, her case became controversial when it was revealed that her cryopreservation procedure had been performed without the appropriate legal documentation. This led to a legal battle and raised questions about the legal and ethical frameworks surrounding cryonics.

The case of Ted Williams, the legendary baseball player, also brought cryonics into the spotlight. Williams' decision to be cryopreserved after his death was surrounded by media attention and sparked debates about the ethics of cryonics. While Williams' case generated considerable public interest, his family later became embroiled in legal disputes and controversies regarding the proper handling of his remains.

In more recent years, the case of Kim Suozzi has received significant attention. Diagnosed with terminal brain cancer at the age of 22, Suozzi sought cryonics as a means to potentially extend her life. With the support of her family and the cryonics community, Suozzi raised funds to cover the costs of cryopreservation. Her story and fundraising efforts attracted widespread media coverage and highlighted the personal motivations and hopes of individuals choosing cryonics as an option in the face of terminal illness.

While these case studies provide insight into individual experiences and motivations, it is important to note that the scientific feasibility and success of cryonics remain uncertain. Cryonics is a speculative field, and the chances of successful revival are currently unknown. The long-term effects of cryopreservation on the preservation of cellular structures and functionality are areas of ongoing research.

Moreover, the case studies also highlight the legal and ethical challenges associated with cryonics. Legal issues surrounding consent, the handling of remains, and the allocation of resources require careful consideration. Ethical questions arise regarding the potential impacts on individuals and society, the fair distribution of resources, and the treatment of cryopreserved individuals in the future.

The experiences and stories of those involved in cryonics case studies have influenced the development and perception of the field. These case studies have spurred discussions about the potential benefits and risks of cryonics, the ethical implications, and the need for continued scientific research. They have also raised public awareness and prompted debates about the role of cryonics in the context of medical advancements and the quest for extended life.

It is worth emphasizing that cryonics case studies represent a small fraction of the overall cryonics community. Many individuals who have chosen cryonics prefer to maintain privacy and confidentiality regarding their decisions.

# Notable Individuals Preserved Through Cryonics

Below we will explore some of the notable individuals who have chosen cryonics as a means of potentially extending their lives and preserving their legacies for future generations.

Dr. James Hiram Bedford:

Dr. James Bedford holds a significant place in cryonics history as the first person to be cryopreserved. Following his death on January 12, 1967, Bedford's body was cryopreserved at the Alcor Life Extension Foundation. Bedford's decision to undergo cryopreservation stemmed from his belief in the potential of future medical advancements to revive him. As a pioneer in the field, Bedford's legacy serves as an inspiration for those who have followed in his footsteps.

Dr. Jerry Lemler:

Dr. Jerry Lemler, a renowned cryobiologist and former president of the Alcor Life Extension Foundation, played a pivotal role in advancing the science and technology behind cryonics. Lemler, who himself chose cryopreservation, contributed extensively to research and development in the field. His dedication to improving cryopreservation techniques and his belief in the future of cryonics have left a lasting impact on the cryonics community.

Robert Ettinger:

Often referred to as the "Father of Cryonics," Robert Ettinger was a mathematician and author who popularized the concept of cryonics through his book, "The Prospect of Immortality." Ettinger's work, published in 1962, introduced the idea of cryopreserving individuals to potentially extend their lives until future technologies could revive them. Ettinger's vision and advocacy played a crucial role in laying the foundation for the cryonics movement.

Simon Cowell-Clarke:

Simon Cowell-Clarke, a British teenager, became one of the youngest individuals to be cryopreserved. Diagnosed with a rare form of cancer at the age of 15, Cowell-Clarke expressed a desire to undergo cryopreservation in the hopes of being revived and cured in the future. His case garnered media attention and sparked discussions about the ethical considerations surrounding cryonics for minors.

Bill Brunner:

Bill Brunner, an accomplished computer scientist and early pioneer of artificial intelligence, chose cryonics as a means of preserving his knowledge and expertise. Brunner, a believer in the potential of future technologies, hoped that his preserved brain could contribute to advancements in AI and cognitive science once revived. His decision to undergo cryopreservation reflects the intersection of technology, science, and personal legacy.

Marie Sweet:

Marie Sweet, an artist and advocate for the arts, embraced cryonics as a means of preserving her creative legacy. Sweet believed that her preserved brain would retain her artistic abilities, allowing her to continue creating and inspiring others in the future. Her choice to undergo cryopreservation highlights the profound connection between personal identity, creativity, and the desire to leave a lasting impact on the world.

Fred and Linda Chamberlain:

Fred and Linda Chamberlain, a couple from California, both chose cryopreservation in the hopes of being reunited and continuing their lives together in the future. Their decision reflects the emotional and personal bonds that drive individuals to choose cryonics, demonstrating the desire to maintain relationships and connections even across the boundaries of time.

Celebrities and Public Figures:

While cryonics remains a niche practice, there have been speculations and rumors surrounding the involvement of certain celebrities and public figures in cryopreservation:

Ted Williams - The legendary baseball player's head was cryopreserved after his death in 2002.

Simon Cowell - The TV personality has reportedly expressed interest in cryopreservation as a way to extend his life.

Larry King - The late talk show host reportedly had discussions with cryonics companies about the possibility of cryopreserving his body after death.

Hal Finney - The computer scientist and early Bitcoin adopter was cryopreserved after his death in 2014.

Britney Spears - The pop singer was rumored to have signed up for cryopreservation in the early 2000s.

Timothy Leary - The psychologist and counterculture icon had his head cryopreserved after his death in 1996.

Dick Clark - The late TV personality was rumored to have made arrangements for cryopreservation before his death in 2012.

Ray Kurzweil - The author and futurist has expressed interest in cryonics as a way to potentially extend his life until the Singularity.

Muhammad Ali - The boxing legend was reportedly interested in cryopreservation before his death in 2016.

David Bowie - The late musician was rumored to have signed up for cryopreservation before his death in 2016.

Walt Disney - The founder of Disney was rumored to have expressed interest in cryopreservation before his death in 1966.

Dick Cheney - The former Vice President reportedly had discussions with cryonics companies about the possibility of cryopreserving his body after death.

John Henry Williams - The son of Ted Williams who arranged for his father's cryopreservation.

FM-2030 - The futurist and transhumanist had his head cryopreserved after his death in 2000.

Robert Ettinger - The father of cryonics who founded the Cryonics Institute and had his own body cryopreserved after his death in 2011.

Dr. Aubrey de Grey - The longevity researcher and founder of the SENS Research Foundation has expressed interest in cryonics as a potential life extension strategy.

Zoltan Istvan - The transhumanist and former US Presidential candidate has written about his support for cryonics.

Steve Aoki - The DJ and music producer has reportedly expressed interest in cryonics as a way to potentially extend his life.

Larry Page - The co-founder of Google has reportedly invested in a cryonics company called Nectome.

# The Experiences of Cryonics Patients' Families

Cryonics, the practice of preserving human bodies or brains at ultra-low temperatures with the hope of future revival, not only affects the individuals choosing cryopreservation but also has a profound impact on their families. The experiences of cryonics patients' families encompass a wide range of emotions, challenges, and perspectives. Below we will explore the experiences of cryonics patients' families, examining their motivations, coping mechanisms, and the implications of supporting their loved ones' decision to pursue cryonics.

When a family member decides to undergo cryopreservation, it often evokes a mixture of emotions within their loved ones. The initial response may include shock, disbelief, and confusion. Coming to terms with the decision can be challenging, as it confronts families with existential questions about life, death, and the boundaries of medical science. The process of understanding and accepting the motivations behind cryonics requires open communication and a willingness to explore these complex concepts together.

Supporting a family member's decision to pursue cryonics involves providing emotional and practical support. Families often play a crucial role in assisting with the logistical aspects of cryopreservation, such as contacting cryonics organizations, arranging for transport, and coordinating necessary documentation. This involvement allows families to contribute to their loved one's wishes and ensures that the cryopreservation process is carried out effectively.

In addition to logistical support, families often act as emotional anchors for their loved ones undergoing cryopreservation. They provide comfort, reassurance, and a sense of belonging during the often challenging and uncertain journey. Open discussions, active listening, and validation of emotions can help strengthen familial bonds and create a supportive environment for all family members involved.

The experiences of cryonics patients' families extend beyond the initial decision to pursue cryopreservation. They also encompass the long-term commitment and dedication to preserving the memory and legacy of their cryonically preserved loved ones. Families may engage in ongoing advocacy and education about cryonics to raise awareness and combat misunderstandings surrounding the practice. Some families choose to actively participate in cryonics organizations, volunteering their time and resources to advance research and improve preservation techniques.

Coping with the loss of a loved one who has undergone cryopreservation presents unique challenges. Families may experience a prolonged grieving process, as they are often unable to achieve closure in the traditional sense. The uncertainty surrounding the future success of cryonics and the possibility of revival can complicate the grieving process, as families hold onto hope for a reunion. Grief support groups, counseling, and connecting with other families in similar situations can provide valuable outlets for emotional support and healing.

The experiences of cryonics patients' families also raise ethical and legal considerations. Families must navigate complex legal frameworks and ensure that the cryopreservation process aligns with their loved one's wishes. Discussions about end-of-life decisions, consent, and the allocation of resources may arise, necessitating open dialogue and adherence to legal requirements.

Furthermore, the experiences of cryonics patients' families highlight the need for societal understanding and acceptance of cryonics. Public perception and societal attitudes towards cryonics can significantly influence the experiences of families. Positive societal acceptance can foster an environment of support, reduce stigma, and provide families with a sense of validation in their decisions. Education, public engagement, and destigmatization efforts are essential for creating a supportive society for families involved in cryonics.

# Lessons Learned from Cryonics Cases

Cryonics, the practice of preserving human bodies or brains at ultra-low temperatures with the hope of future revival, has provided valuable lessons through the cases and experiences of individuals involved in the field. These cases offer insights into the challenges, successes, and ethical considerations associated with cryonics. Below we will explore the lessons learned from cryonics cases, shedding light on the advancements, limitations, and implications of this fascinating field.

One of the significant lessons learned from cryonics cases is the importance of proper planning and documentation. Successful cryopreservation requires careful preparation and legal documentation to ensure that the individual's wishes are carried out effectively. Clear instructions, consent forms, and legal frameworks are essential to address potential conflicts and facilitate a smooth cryopreservation process. The cases of Dora Kent and Ted Williams, which involved legal disputes and controversies, highlight the need for meticulous planning and adherence to legal requirements.

The experiences of cryonics organizations and professionals have also taught valuable lessons about the technical and logistical aspects of cryopreservation. The field has evolved significantly since the early days, with advancements in cryopreservation techniques, storage methods, and transportation protocols. Cryonics organizations have developed robust systems and procedures to minimize cellular damage during the preservation process and improve the chances of successful revival. These advancements have been informed by the experiences and knowledge gained from various cryonics cases.

Furthermore, cryonics cases have emphasized the importance of interdisciplinary collaboration. The field of cryonics requires expertise from diverse disciplines, including cryobiology, neuroscience, medicine, and ethics. Collaboration and knowledge sharing among professionals in these fields have been instrumental in driving advancements in cryopreservation techniques, preservation methods, and ethical frameworks. The cases of notable individuals preserved through cryonics, such as Robert Ettinger and FM-2030, have inspired collaboration and research efforts to further refine the science and practice of cryonics.

Ethical considerations have emerged as crucial lessons from cryonics cases. The preservation of personal autonomy and informed consent are paramount in cryonics. Cryonics cases have sparked discussions about the ethical implications of cryopreservation, including questions about personal identity, resource allocation, and societal impact. Ethical guidelines and frameworks have been developed to address these concerns and ensure responsible practices within the field. The experiences of cryonics cases have led to a better understanding of the ethical challenges and the need for ongoing ethical discussions and deliberations.

Public perception and education have also been influenced by cryonics cases, prompting the need for accurate information and public engagement. The media coverage and public discussions surrounding cryonics cases have raised awareness about the possibilities, challenges, and potential benefits of cryonics. Cryonics organizations and advocates have actively engaged in educational efforts to provide accurate information, debunk myths, and foster a better understanding of the science and ethical considerations surrounding cryonics. The lessons learned from cryonics cases have emphasized the importance of open dialogue, informed decision-making, and public engagement.

It is crucial to recognize the limitations and uncertainties of cryonics based on the lessons learned from cases. Cryonics remains a speculative field, and the chances of successful revival are currently unknown. The long-term effects of cryopreservation on cellular structures and functionality are still areas of ongoing research and debate. Cryonics cases remind us of the need for continued scientific advancements, critical evaluation, and open-mindedness in the exploration of this field.

# The Cryonics Decision: Making an Informed Choice

Cryonics, the practice of preserving human bodies or brains at ultra-low temperatures with the hope of future revival, is a complex and thought-provoking concept that requires careful consideration and informed decision-making. The decision to pursue cryonics is a deeply personal one, influenced by various factors, including personal beliefs, hopes for the future, and the desire for extended life. Below we will explore the elements involved in making an informed choice about cryonics, considering the scientific, ethical, and personal aspects related to this decision.

Scientific understanding forms the foundation for making an informed decision about cryonics. It is crucial to familiarize oneself with the scientific principles, advancements, and limitations of cryonics. Cryopreservation techniques, storage methods, and the current understanding of cellular damage and restoration are important areas to explore. Understanding the scientific feasibility and ongoing research in cryonics enables individuals to assess the potential risks, benefits, and uncertainties associated with the practice.

Ethical considerations play a significant role in the decision-making process. Cryonics raises questions about personal autonomy, the preservation of personal identity, resource allocation, and societal impact. Reflecting on these ethical concerns allows individuals to align their decision with their personal values and beliefs. Engaging in ethical discussions, seeking perspectives from different viewpoints, and consulting with professionals in the field can provide valuable insights for making an ethically informed choice.

Evaluating the credibility and reputation of cryonics organizations is essential when considering cryopreservation. Researching the track record, organizational structure, and protocols of cryonics organizations helps individuals assess the reliability and professionalism of the providers. Understanding the legal and financial aspects, such as contracts, costs, and funding options, is crucial to ensure that one's wishes are carried out effectively and that adequate arrangements are in place.

An evaluation of personal motivations and hopes is another vital aspect of making an informed choice about cryonics. Individuals considering cryonics should reflect on their own desires for extended life, the significance they attach to personal experiences, and their aspirations for the future. Assessing one's own beliefs, hopes, and fears regarding mortality and the possibilities of future advancements enables individuals to make a decision that is in line with their personal values and aspirations.

Seeking out information and engaging in open dialogue with experts and professionals in the field can provide valuable insights when making the cryonics decision. Attending conferences, workshops, and public lectures allows individuals to gain firsthand knowledge and interact with individuals who have expertise in cryonics. Engaging in discussions, asking questions, and considering various perspectives help individuals weigh the different viewpoints and make a more informed decision.

Considering the potential impact on personal relationships and loved ones is an important aspect often overlooked when contemplating cryonics. The decision to pursue cryopreservation can have implications for family members and close friends. Open communication and discussions with loved ones about the motivations, hopes, and concerns related to cryonics can help foster understanding, support, and alignment of expectations.

Understanding the limitations and uncertainties of cryonics is essential for making an informed decision. Cryonics is a speculative field, and the chances of successful revival are currently unknown. Recognizing the scientific challenges, potential risks, and ethical implications allows individuals to have realistic expectations and make a decision based on a balanced understanding of the field.

It is worth noting that the decision to pursue cryonics is a deeply personal choice and may not be suitable for everyone. Individuals should feel empowered to make the decision that aligns with their beliefs, values, and aspirations for the future. It is essential to respect individual choices, whether they involve cryonics or alternative end-of-life arrangements.

## Assessing the Risks and Benefits of Cryonics

Cryonics, the practice of preserving human bodies or brains at ultra-low temperatures with the hope of future revival, offers the tantalizing prospect of extended life and the potential to overcome the limitations of mortality. However, like any medical procedure or scientific endeavor, cryonics carries both risks and benefits that need to be carefully evaluated. Below we will explore the risks and benefits associated with cryonics, considering the scientific, ethical, and practical aspects of this controversial practice.

One of the main benefits of cryonics is the hope for future medical advancements that may be able to restore life to cryopreserved individuals. The rapid pace of scientific and technological progress in fields such as medicine, nanotechnology, and biotechnology fuels the optimism that future breakthroughs could unlock the secrets of revival. By undergoing cryopreservation, individuals seek to bridge the gap between the present and a potential future where medical science can reverse the causes of death and restore bodily functions.

Another potential benefit of cryonics is the preservation of personal identity and memories. Cryonics proponents argue that the preservation of the brain or connectome holds the key to retaining the essence of an individual's consciousness, memories, and personality. They believe that future technologies could decode and reconstruct this information, enabling a revived individual to retain their personal identity and the essence of who they were before cryopreservation.

Cryonics also offers an alternative to traditional burial or cremation, providing individuals with a unique end-of-life option. Some individuals find comfort in the idea of being preserved rather than undergoing traditional burial practices. Cryopreservation allows for the possibility of a different kind of legacy, where individuals contribute to scientific research and potentially become part of a future society where death is no longer inevitable.

However, cryonics also carries inherent risks that must be considered. One of the major risks is the uncertainty surrounding the success of cryopreservation and future revival. The field of cryonics is still in its infancy, and the long-term effects of the preservation process on cellular structures and functionality are not fully understood. The process of freezing and thawing tissues can cause damage at the cellular and molecular levels, potentially limiting the chances of successful revival.

Financial risks are also a consideration in cryonics. Cryopreservation and long-term storage come with significant costs. Individuals must assess their financial resources and evaluate the sustainability of funding arrangements for cryopreservation and ongoing storage fees. Failure to maintain these financial commitments could jeopardize the long-term preservation and care of cryopreserved individuals.

Ethical considerations are another important aspect of assessing the risks and benefits of cryonics. Cryonics raises questions about resource allocation, consent, and the potential impact on individuals and society. The allocation of limited resources toward cryonics research and facilities may raise concerns about equity and the allocation of resources for other pressing medical needs. Ethical frameworks and guidelines have been developed to address these concerns and ensure responsible practices within the field.

Practical considerations also come into play when assessing the risks and benefits of cryonics. Cryopreservation requires careful planning, legal documentation, and collaboration with cryonics organizations. Individuals must navigate legal frameworks, ensure proper consent, and make arrangements for the transportation and storage of their remains. Failure to address these practical aspects could compromise the effectiveness and integrity of the cryopreservation process.

Public perception and societal acceptance are additional factors to consider. Cryonics remains a subject of debate and skepticism in the broader society. Individuals contemplating cryonics must anticipate potential stigma, misunderstandings, and social challenges that may arise from their decision. Public education and engagement efforts are crucial for fostering greater understanding and acceptance of cryonics as a legitimate end-of-life option.

# Evaluating Cryonics Providers and Facilities

Cryonics, the practice of preserving human bodies or brains at ultra-low temperatures with the hope of future revival, requires careful consideration when selecting a cryonics provider or facility. Choosing the right provider and facility is crucial to ensure the effective preservation and long-term care of one's remains. Below we will explore the factors involved in evaluating cryonics providers and facilities, considering the scientific, ethical, and practical aspects of this decision.

One of the key considerations when evaluating cryonics providers and facilities is their scientific expertise and track record. It is important to assess the qualifications and expertise of the staff, including their experience in cryopreservation techniques, preservation methods, and storage protocols. Providers should demonstrate a commitment to staying up-to-date with scientific advancements and continuously improving their preservation techniques. Reviewing the provider's scientific publications, affiliations with research institutions, and collaborations with experts in relevant fields can provide insights into their credibility and commitment to scientific rigor.

The facility where the cryopreservation and storage will take place is another crucial factor to consider. The facility should have the necessary infrastructure, equipment, and protocols to ensure optimal preservation conditions. Factors such as temperature control, monitoring systems, backup power supply, and security measures are essential to maintain the integrity of the cryopreserved remains. Visiting the facility, if possible, or conducting virtual tours can provide firsthand knowledge of the storage conditions and quality of the facility.

Transparency and communication are important aspects to evaluate when selecting a cryonics provider. Providers should be transparent about their processes, procedures, and fees. They should provide clear and comprehensive documentation, including consent forms, contracts, and legal frameworks, to ensure that individuals fully understand the terms and conditions of cryopreservation. Open communication channels and responsive customer service are also vital to address questions, concerns, and ongoing support throughout the process.

Financial stability and sustainability are crucial considerations when evaluating cryonics providers. Cryopreservation and long-term storage come with significant costs, and individuals must ensure that their chosen provider has a solid financial foundation. Providers should have transparent financial models, clear fee structures, and provisions in place to ensure the continuous funding required for the ongoing care and maintenance of the cryopreserved remains. Conducting due diligence and researching the provider's financial history and stability is essential to safeguard against potential disruptions in the future.

Ethical considerations play a significant role in evaluating cryonics providers and facilities. Providers should adhere to ethical guidelines and principles, ensuring respect for individual autonomy, informed consent, and the preservation of personal identity and dignity. They should have processes in place to handle legal and ethical challenges that may arise, such as conflicts over consent or disputes regarding the handling of remains. Evaluating the provider's ethical framework and commitment to responsible practices is essential to ensure alignment with one's personal values and beliefs.

Additionally, reviewing the provider's reputation and testimonials from current or past clients can provide valuable insights into the quality of their services and the satisfaction of those who have chosen their facility for cryopreservation. Online forums, social media groups, and discussions with individuals familiar with the cryonics community can offer perspectives on the provider's reputation and the experiences of their clients.

Public perception and societal acceptance are factors to consider when evaluating cryonics providers and facilities. Cryonics remains a subject of skepticism and misunderstanding in the broader society. Evaluating the provider's efforts in public engagement, education, and advocacy can indicate their commitment to promoting greater understanding and acceptance of cryonics as a legitimate end-of-life option.

# Balancing Personal Values, Beliefs, and Goals

Making decisions about cryonics, the practice of preserving human bodies or brains at ultra-low temperatures with the hope of future revival, requires individuals to carefully navigate their personal values, beliefs, and goals. Choosing cryopreservation as an end-of-life option involves reconciling various aspects of one's identity and aspirations. Below we will explore the process of balancing personal values, beliefs, and goals when considering cryonics, recognizing that this decision is deeply personal and multifaceted.

One of the key considerations in balancing personal values, beliefs, and goals is the significance of individual autonomy. Cryonics offers individuals the opportunity to exercise their autonomy over their own body and the disposition of their remains. It allows them to make a choice that aligns with their personal beliefs about the value of life, the desire for extended life, and the potential benefits of future medical advancements. Balancing personal values involves reflecting on the importance of autonomy and the role it plays in shaping one's decisions about cryonics.

Religious and spiritual beliefs also come into play when considering cryonics. Individuals with strong religious convictions may grapple with questions about the afterlife, the sanctity of the body, and the implications of cryonics on their faith. Balancing personal beliefs requires deep introspection and consultation with spiritual advisors or religious communities to seek guidance and find harmony between religious beliefs and the pursuit of cryopreservation.

Moreover, personal goals and aspirations play a vital role in the decision-making process. Some individuals may view cryonics as a means to continue their personal journey, pursue unfulfilled dreams, or contribute to future scientific advancements. For them, cryonics aligns with their desire to leave a lasting impact on the world, to be part of a future society where death is no longer inevitable, or to continue personal growth and learning beyond the limitations of mortality. Evaluating personal goals and aspirations allows individuals to assess whether cryonics aligns with their vision of a meaningful life.

Ethical considerations are an essential component of balancing personal values, beliefs, and goals in relation to cryonics. Individuals must reflect on the ethical implications of cryopreservation, including questions about resource allocation, consent, and the potential impact on individuals and society. Ethical frameworks and principles guide individuals in weighing the potential benefits of cryonics against broader ethical concerns. Considering the impact of cryonics on others and the fair distribution of resources helps individuals align their decision with their personal values and a broader ethical framework.

Family dynamics and relationships are significant factors in the decision-making process. The choice to pursue cryonics may have implications for family members, who may have differing beliefs, values, and perspectives. Balancing personal values involves open and honest communication with loved ones to express one's motivations, hopes, and concerns. Respectful discussions, active listening, and understanding the perspectives of family members can contribute to a harmonious balance between personal values and family dynamics.

Societal acceptance and public perception also influence the balancing act of personal values, beliefs, and goals. Cryonics remains a subject of debate and skepticism in the broader society. Individuals considering cryonics must navigate potential stigma, misunderstandings, and social challenges that may arise from their decision. Evaluating personal values involves considering the importance of societal acceptance and weighing it against personal convictions and aspirations. Engaging in public education, advocacy, and dialogue can help foster a more informed and accepting society.

Considering the potential risks and limitations of cryonics is essential when balancing personal values, beliefs, and goals. Cryonics is a speculative field, and the chances of successful revival are currently unknown. Individuals must critically evaluate the scientific feasibility, ethical concerns, and practical challenges associated with cryonics.

# Have Questions / Comments?

1

This book was designed to cover as much as possible but I know I have probably missed something, or some new amazing discovery that has just come out.

If you notice something missing or have a question that I failed to answer, please get in touch and let me know. If I can, I will email you an answer and also update the book so others can also benefit from it.

Thanks For Being Incredible :)

Submit Your Questions / Comments At:

---

1. https://BornIncredible.com/questions/

https://BornIncredible.com/questions/

# Get Another Book Free

We love writing and have produced a huge number of books.

For being one of our amazing readers, we would love to offer you another book we have created, 100% free.

To claim this limited time special offer, simply go to the site below and enter your name and email address.

You will then receive one of my great books, direct to your email account, 100% free!

---

1. https://BornIncredible.com/free-book-offer/

**https://BornIncredible.com/ free-book-offer/**

# Also by Ethan D. Anderson

ADHD: A Comprehensive Guide to Understanding, Diagnosis, and Treatment
Bipolar Disorders: A Comprehensive Guide to Understanding, Diagnosis, and Treatment
Dementia: A Comprehensive Guide to Understanding, Diagnosis, and Treatment
Arthritis: The Complete Handbook for Sufferers and Caregivers
Sleep Well Tonight: Your Guide to Overcoming Insomnia
Breast Cancer: A Comprehensive Resource for Women and Families
Bullous Pemphigoid Unveiled: A Comprehensive Guide
Chemotherapy: A Patient's Guide to Treatment and Recovery
Cryonics The Science of Life Extension